AF363076

Third Edition

GUTTA PERCHA

A Journey

With Appendices on
Xylonite (Celluloid)/Amber & Copal

Masatoshi Iguchi

Texnai

Texnai®
3-5-24-406 Nakanoshima, Tama-ku, Kawasaki,
Kanagawa 2140012, Japan
Tel: 081-863-9545
Fax: 081-863-9597

ISBN 978-4-909601-84-1

Published 31 January 2021
First print 31 January 2021
Second Edition 25 March 2021
Third Edition 15 October 2021

Typeset by the author

Printed by KDP Amazon.com. Inc.

Foreword by Prof. Malcolm R. Mackley

Gutta Percha has a special place in the history of plastics and rubbers and the author of this review captures the very soul of this fascinating material. Gutta Percha is obtained from the latex of Sapotaceae family trees indigenous to Southeast Asia and in the 1850s it was discovered that if the latex was processed in a particular way, the material could be melted and then moulded into solid complex shapes. This marked the beginning of a new age of polymer plastics made initially from biopolymers and then later in the twentieth century from synthetic polymers. Gutta Percha is thus the first thermoplastic polymer to be discovered and is the precursor to a class of material that has transformed our lives in so many different ways.

Dr Toshi Iguchi is a distinguished polymer physicist who has had extensive scientific experience of polymeric material and who also has a fascination with naturally occurring polymers from areas such as Indonesia. In this review he combines his knowledge of polymers and Indonesia to create a very personal account of the way the material and technology has developed over more than 150 years. The journey will take the reader into the heart of Indonesian jungle in order to discover Gutta Percha processing plants that by 21st standards are of brilliant eco-design, where power for example was generated from waste coconut shells. This story should fascinate anyone interested in the history of material science and also Indonesia.

Professor Malcolm R. Mackley
Emeritus Professor of Chemical Engineering,
University of Cambridge,
UK.
December 2020

Preface

Only people of the present writer's age or older has experienced the time when virtually no synthetic plastics existed around them. Xylonite, or Celluloid, which was of use for toys, stationeries, glass-frames and so on, was the mixture of nitrocellulose and camphor heated at elevated temperature. The substrate of photographic- and cinematographic films was made of cellulose acetate. Bakelite used for electric parts and eating utensils was obtained from the mixture of phenol and formaldehyde cured at high temperature, thus classified as thermoset resin. Natural rubber was used for various purposes ranging from pencil eraser to pneumatic tyres, whether it should be included in the category of plastics. Nylon, or *poly-hexamethylene-diamine*, had been invented in 1935, but the primary aim of this polymer was to manufacture textile fibres. Polythene, also discovered in the 1930s, was yet to be a commodity polymer.

Gutta percha, originated from the latex of Sapotaceae trees, had held a special status as the only thermoplastic polymer until the 1970s when the applications were superseded by synthetic polyolefins.

At school (in 1950s) the present writer was taught that hydrofluoric acid must be kept in a bottle of gutta percha which was obtained from a tropical plant, because the acid corroded glass. The bottle he saw appeared to be candle wax but was hard and unscratchable by a nail. He was also taught that gutta percha was indispensable for the insulative covering of submarine telephone cables that connected all continents and islands in the world. At university he learnt that the polymer was *trans-1,4-poly-isoprene,* which, by virtue of the specific configuration, stayed solid below ca. 60°C, unlike its isomer, *cis-1,4-poly-isoprene,* or natural rubber.

Originally utilised by native people of Malay Archipelago for tools and artefacts, gutta percha was first brought to Europe in 1843. The history after then, including the founding of the plantation at Tjipetir (Cipetir), was spectacular as will be described in the main part of this booklet.

After the chapters on gutta-percha, two short reviews about other thermoplastic substances will be appended. One is about the nitrocellulose-camphor mixture, i.e., parkesine, xylonite and celluloid, which was useful for photographic and cinematographic films, table tennis balls, toys, and so on, but replaced by synthetic polymers such as polyacrylates and polycarbonate, almost disappeared as readers seldom see them anywhere. The other is about fossilised and semi-fossilised natural resins, amber and copal. The heat-press technique has long been used for fusing small pieces of amber into a single piece and also for modifying defected amber. Copal, which is fluid at elevated temperature, has been used to prepare various moulded objects. It should be noted that nitrocellulose-camphor mixtures, as well as copal, are thermally not fully reversible, as their structure changes to some extent once subjected to high temperature, unlike such synthetic thermoplastic polymers as polyolefins and polyacetals.

The present writer wishes that readers will reconsider the value of gutta percha and other polymers of natural origin and their derivatives, in the circumstances when the dispersion of synthetic plastic waste is a serious world-wide issue, without taking this book as granted as an elegy for gutta percha.

Acknowledgements
The present writer wishes to express his sincere thanks and condolences to the late Dr. Suharto Honggokusumo, the former director of the Bogor Research Station for Rubber Technology, and the late Dr. Augustine Francis Sutrisno Budiman, the former honorary research associate of the same research station, from both of whom he learnt a great deal about gutta percha and natural rubber. He also thanks Prof. Malcolm R. Mackley (Cambridge), Prof. Pieter J. Lemstra (Eindhoven), Dr. Doetze Sikkema (formerly Akzo-Nobel), Prof. Paul Smith (ETH Zürich), Dr. John Goldsbrough (formerly BX Plastics) and Prof. Khoji Tashiro (Toyota Technological University) for their encouragements for writing this article. In particular, Prof. Mackley has kindly edited the text to bring it up to be readable as English and contributed a foreword. Prof. Smith has kindly given advice for the cover design.

Masatoshi Iguchi,
Tokyo, Japan.
E-mail: maiguch@gmail.com
Website: http://www.maiguch.sakura.ne.jp/
December 2020

About the author

Dr. Masatoshi Iguchi, born in 1938 in Nagoya, Japan, had finished the Graduate School of Tokyo Institute of Technology, awarded a doctor of engineering in 1966. He served most of his career time for a governmental research institution under the Ministry of Industrial Science and Technology, Japan, to work in the area of polymer science and technology until retired in 1999. Then, he obtained a fellowship from the Science and Technology Agency, Japan, and worked at Bogor Research Station for Rubber Technology, Bogor, West Java, Indonesia, for three years. After retired from full-time occupations in 2002, he became interested also in the history and culture, viz. of Indonesia.

Notes on the Third Edition

In the Second Edition issued on 5 March 2021, an error in Figure 31, in which the structural formulae of natural rubber and gutta percha had been reversed left and right, and a few typographical errors corrected. In this Third Edition, more errors have been corrected and some parts rewritten as follows.

(1) Page 11, Line 11: A phrase that was missing has been inserted as below,

It was not from his manual abilities <u>alone that his father had given him the</u> name of Gutta-Percha Willie, but from the fact that ...

(2) Page 12, Line 4 from the bottom: Frankfort-on-Maine → <u>Frankfurt</u>-on-<u>Main</u>

(3) The Latin name of balata tree, *Mimusops balata*, in Page 35, Line 2 from the bottom, and Page 64, Line 3, has been corrected as <u>*Manilkara bidentata*</u>.

(4) Page 27, Figure 13: The array of pictures has been changed to match the legend.

(5) Page 58, Figure 43: The legend has been rewritten to match the pictures.

(6) Page 78, Figure 74: The figure has been redrawn on CorelDRAW as similar as possible to clarify the details.

(7) Page 81, Line 7: sulphur → <u>camphor</u>.

(8) Page 85. Figure A1-5: The images on the left- and right-handed sides have been reversed to match the legend.

(9) Page 93, Line 6:

pyroxylin-collodion mixture → <u>camphor</u>-collodion mixture

In addition, some wordings have been modified, some descriptions added and a number of typographic errors amended.

September 2021

Front cover illustration
The oldest gutta-percha tree in Indonesia and probably in the world, *Palaquium oblongifolium* Burck, planted in the 1880s by Dr. W. Burck and his members in the Bogor Botanical Garden.

Inner cover illustration
The Gathering of Gutta Percha. Duplicated from: Green, Benjamin L., Gutta percha, its discovery, history, and manifold uses. Benjamin L. Green, London, 1851.

Rear cover illustration
(1) Native's harvesting and utilisation of gutta percha
    Collection of gutta percha by tapping method; Tool handles; Household utensils.
(2) Introduction to Europe, various applications
    Eating utensils; leather soleing; Gutta percha soleing; Railway conversation tube.
(3) Exclusive applications
    Submarine cable (Dover-Calais, 1851); Denture and dental fillers (1886); Golf balls, Gutty (solid gutta ball, 1848) and Haskell (rubber ball covered with balata, 1899).
(4) Large-scale plantation with nursery and modern factory at Tjipetir, West Java
    A view of the Tjipetir factory; Harvested Palaquium oblongifolium leaves; Cooking and separating process. (1920-1930);
(5) The India Rubber, Gutta Percha and Telegraph Works Company, Silvertown
    The Silvertown Telegraph Cable Works is seen at the right-end (ca. 1869).

Contents

Introduction

In the beginning paragraph of his novel, *The History of Gutta Percha Willie, the Working Genius (1873)*,[1] George MacDonald, the prominent Victorian author, poet and minister, put a premise,

> *His name in full was Willie Macmichael... It was his own father, however, who gave him the name of Gutta-Percha Willie, the reason of which will also show itself by and by.*

and gave the reason in a later paragraph, as,

> *It was not from his manual abilities alone that his father had given him the name of Gutta-Percha Willie, but from the fact that his mind, once warmed to interest, could accommodate itself to the peculiarities of any science, just as the gutta-percha which is used for taking a mould fits itself to the outs and ins of any figure.*

So the word, gutta percha must have been in common use among a contemporary audience of that time.

Today's young readers are probably not very familiar with gutta percha as gutta percha-covered golf balls had almost completely disappeared around the turn of the century to the year 2000 and no products of this substance other than dental fillers are now found. In fact, the present writer has checked and realised that in most chemistry textbooks published in recent decades the description on gutta percha is rather short, such that gutta percha is *trans*-1,4-poly-isoprene obtained from some tropical plants and that, by virtue of the specific configuration, the polymer is solid below ca. 60°C, unlike its isomer, *cis*-1,4-poly-isoprene or natural rubber, without mentioning the fact that, prior to the emergence of synthetic substitutes, gutta percha was the only thermoplastic material of technical importance, indispensable especially for the insulative covering of submarine cables, the container of hydrofluoric acid, anticorrosive gaskets, etc.

Gutta percha (*lit.* gum of tree) was known to occur in the bark of some trees of Sapotaceae family indigenous to Malaya and Malay

Archipelago. Native people were aware of the peculiar property that the substance easily solidifies when exuded from trees, unlike the latex of natural rubber, but softened in warm water. They traditionally made use of this substance, locally also called *Getah Taban Sutra*, *Getah Taban Merah*, etc., for the handle of tools, various household utensils and artefacts.

The history of this unique substance after its existence became known to Europeans in the early nineteenth century just in the height of the industrial revolution was spectacular. Many entrepreneurs invested capital to acquire and apply this material for various purposes. Within a matter of a few decades, special machineries to process this first "thermoplastics" was developed, many enterprises, including The Gutta Percha Company, were established and a variety of products were entered in the Great Exhibition of the Works of Industry of All Nations 1851, London. Two books on the "history" and usefulness of gutta percha were published. The title of the first book, *Gutta-Percha, Its Discovery, History, Remarkable Properties, Vast Utility, and Application to Scientific and Ornamental Purposes, also Its Economy, and Importance as a Sanatory agent, with Instructions for Soling Boots and Shoes, Joining Driving Bands and Tubings, and for the Preventing Toothache, by Stopping the Caries* (1849),[2] shows how people were enthusiastic at that time. The second book, *Gutta percha, its discovery, history, and manifold uses* (1851)[3] depicted the contemporary processes as well as all sort of products.

In the meantime, the fact that the insulating property of gutta percha was as good as that of shellac[4] was found by Michael Faraday in mid 1840s.[5] In the same decade, Werner Siemens started to apply gutta percha for the insulative covering of cables and his cable was laid under the soil between Berlin and Potsdam, Berlin and Frankfurt-on-Main, etc.[6] The first submarine cable across the English Channel between Dover and Calais was laid in 1851. Then before the early twentieth century almost all continents and islands around the world were linked by telegraph cables.

It should be noted that a contemporary chemist, Greville Williams, had raised an idea in 1860 that gutta percha as well as caoutchouc (natural rubber) was a *polymeric body* which could be disrupted by the action of heat into a substance, *isoprene*, having a simple relation to the parent hydrocarbon, the chemical formula of which being C_5H_8.[7] It was long before the so-called Macromolecular Theory was proposed and accepted in the 1920s. For reference, the fact that isoprene polymerised into "India-rubber" when stored in a bottle, presumably by the action of a small amount of some acidic contaminants or the ray of light, was reported by William A. Tilden in 1892.[8]

With the increased demand for gutta percha, trees in many forests in the Island of Singapore, Malay Peninsula and elsewhere rapidly diminished. To cope with this problem, the cultivation of the tree was attempted in 1848 and subsequent years in Singapore but the result was not successful.[9; 10] The genus of the tree of use at that time was mainly *Isonandra* of the Sapotaceae family.

It was at the Buitenzorg Botanical Garden (the present Bogor Botanical Garden), West Java, that source plants as well as efficient methods to exploit gutta percha were seriously studied. In 1884, Dr. William Burck collected *Palaquium oblongifolium* and several other gutta percha-producing plants (*Palaquium gutta, Palaquium Treubii, Palaquium Borneense, and Payena Leera*) and planted them in the Garden,[11] as highlighted in the guidebook authored by himself.[12] In 1885, a vast Experimental Garden was opened by the Dutch Government at Tjipetir, Preangan Regency (the present-day Cipetir, Kabupaten Sukabumi) and a large-scale project to develop forests, started. "*Palaquium oblongifolium*, Burck", which had been determined to be the best of all, bore abundant fruits in 1895.[13] The grafting which was said to be impossible in the past was successfully done. The method to obtain high-quality gutta percha (leaf-gutta, also called white gutta) from harvested leaves and twiglets was established and a factory equipped with modern machines and chemical apparatus, constructed.

The present writer visited Cipetir for the first time in 1992 and admired the forest and factory which was maintained in operational conditions, despite the fact that the amount of production had been drastically reduced to the level of a few tons per year. The pile of white gutta (pure *trans*-1,4-poly-isoprene) moulded in the form of a disk ready to be packed for shipping was very impressive. When the idea to hold an international workshop on green polymers in Java (1996) was proposed by Prof. P. J. Lemstra, Eindhoven University of Technology, and the present writer was involved in its planning, the latter recommended the Cipetir Plantation as one of the sites of the study tour so that the participants could see the real specimen as well as the forest and production process of the once very important polymer.

During his stay in the Bogor Research Station for Rubber Technology (now Bogor Rubber Research Centre, Indonesian Rubber Research Centre, a laboratory originally branched in 1915 from the Botanical Garden with Dr. Otto de Vries as the first director), as a research fellow in 1999-2002, the present writer had heavy discussion on the past and future of gutta percha with Dr. Suharto Honggokusumo, the then director, and Dr. Augustine Francis Sutrisno Budiman, the then an honorary research associate, who were the last two persons who had associated with gutta percha research.

Now that both Dr. Suharto and Dr. Budiman have passed away, the present writer has felt as if it was his duty to review the glorious history of gutta percha with modern knowledge of polymer science. It was incidental that last year Prof. Lemstra told his close friends of his idea to write a book on *the rise and decline of plastics*, and in response to his order the present writer had made a survey on gutta percha as well as xylonite (called celluloid in America) and ebonite (highly vulcanised natural rubber) which were also important in the past. In this article, the result on gutta percha will be summarised tracing as much original articles as possible and adding photo-images and illustrations.

Chapter 1
The source of gutta percha and the use by native people

As mentioned above, gutta percha was known to occur in the bark of trees of the Sapotaceae family, viz. the genera of *Isonandra* and *Palaquium*, which grow, or grew in the past, in the tropical forests in Southeast Asia. Figure 1 shows the botanical illustration of *Isonandra gutta* and *Palaquium oblongifolium*, Burck and Figure 2, the same of *Hevea brasiliensis*, for reference.

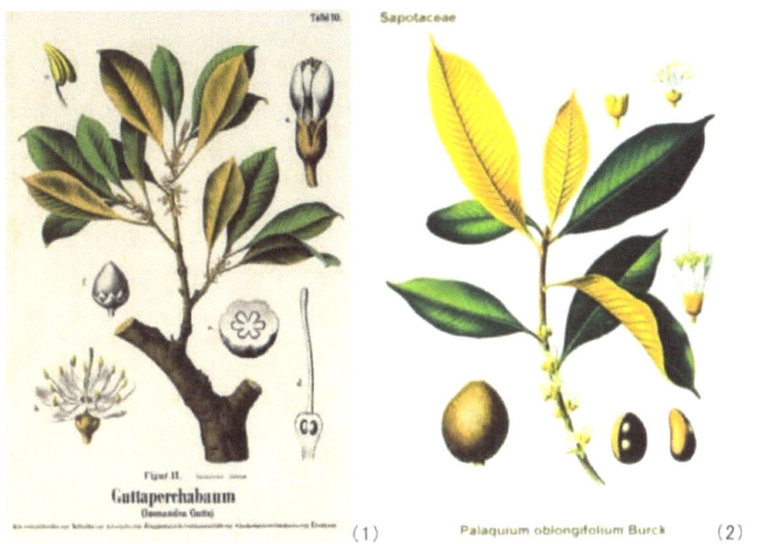

Figure 1 Botanical illustration of *Isonandra gutta* (1)[14] and *Palaquium oblongifolium*, Burck (2).[15]

Figure 2 Botanical illustration of *Hevea brasiliensis*.[16]

Native people traditionally obtained gutta percha from the bark of trees felled down in forests. After chopping off the branches, they ringed the bark at distances of 12 to 18 inches all along the trunk, whereupon the latex or the milky sap soon exuded and filled the grooves cut into the bark and quickly coagulated. They collected the solid gutta percha by means of a knife or a scraper and manually removed dirt in warm water.

Although the tapping method was introduced to save the life of trees, the method was not efficient either, because of the fact that the latex receptacles existed in the bark as isolated, discontinuous bodies (cf. Figure 3 and Figure 4) and that the latex was thicker and less fluidly than that of the natural rubber.

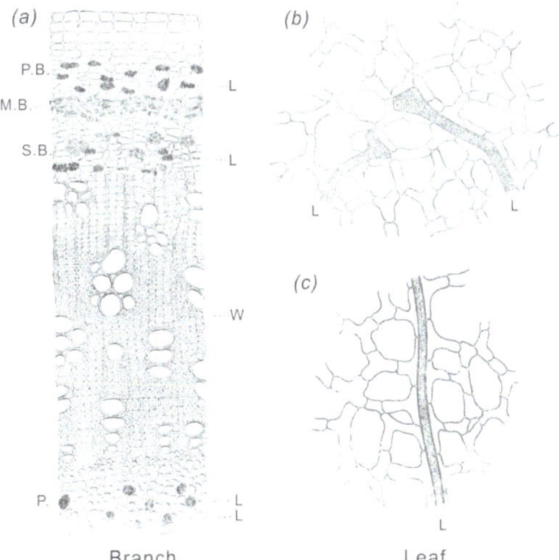

Figure 3 Microscopic section of *Palaquium Gutta* branch of 6 mm diameter. P.B.: Primary bark, M.B.: Mixed band, S.B.: Secondary Bark, P: Pith, W: Wood body, L: Latex sac.[17]

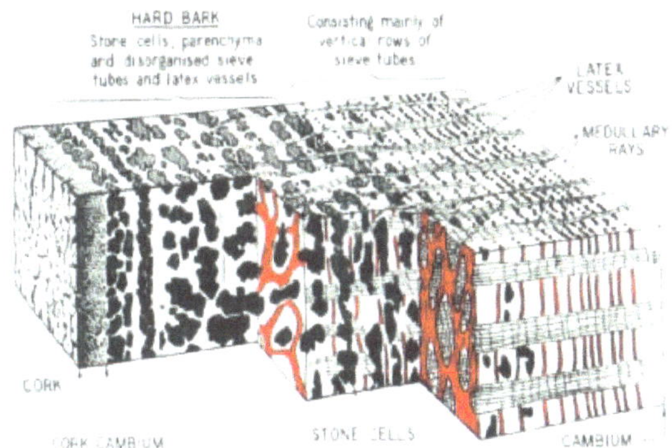

Figure 4 Three dimensional diagram of the bark of *Hevea brasiliensis*.[18]

Figure 5 illustrates the scenes of collecting gutta percha from trees in the forest, found in old literature.

Figure 5 Collecting of gutta percha in forest. (1) Extraction of gutta percha from the trunk of fell-down trees in Sumatra[19] (2) Four Kayan natives collecting gutta percha from a tree trunk in Sarawak, c. 1896.[20] (3) Collection of gutta percha in Malaya[21] (4) Removing naturally coagulated gutta-gum from freshly opened channels. Traat, eastern Thailand.[22]

Figure 6 shows old photographs of gutta percha products made and used by native people.

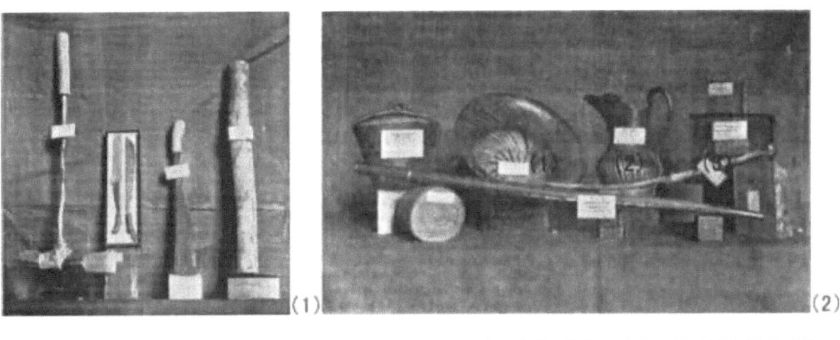

Figure 6 Native people's gutta percha products. (1)Tool handles, (2) Household utensils and (3) Artefacts.[23]

Even today, they make use of gutta percha for souvenirs, such as shown in Figure 7, as the present writer witnessed and bought some pieces at the shop of Mulawarman Museum, Samarinda, East Kalimantan. He wished to visit their home to see how they get and process the material but unfortunately he was told by the shopkeeper that "Countryside people just occasionally come to deliver them. We can ask the place of their home next time". He was certainly unable to extend his stay in Samarinda for uncertain days or weeks.

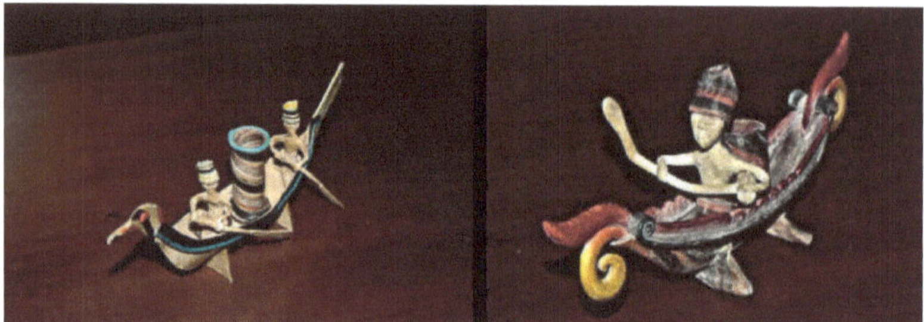

Figure 7 Souvenirs from the shop of Museum Mulawarman, Samarinda, Kalimantan (Photographed by the present writer, 2007).

At the Samarinda Airport Art Shop, he found and purchased a model of "Buginese Ship", made all of gutta percha and coloured with natural pigments, as seen in Figure 8, which was highly artistic and the best of this kind he had ever seen. His colleagues in Bogor, especially Dr. Budiman, appreciated this beautiful piece.

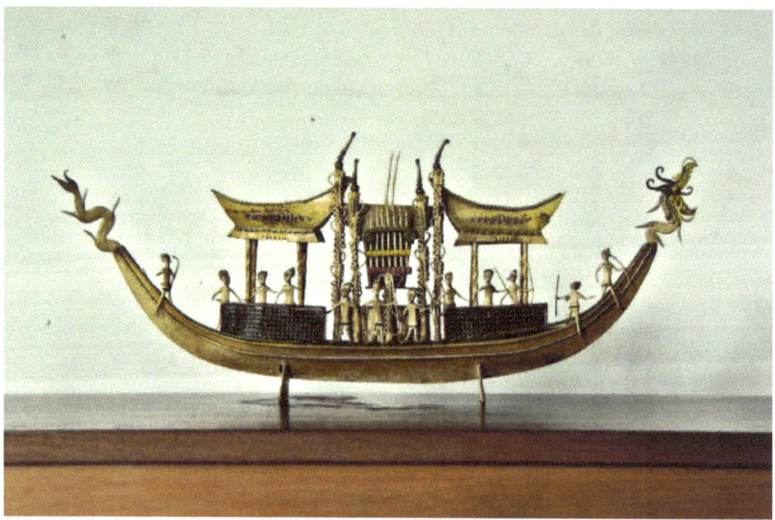

Figure 8 Model of "Buginese Ship", purchased by the present writer at the Samarinda Airport Art Shop and housed in the Bogor Research Station for Rubber Technology (Photographed by the present writer, 2007).

As regards the amount of solid gutta percha yielded by an adult tree, the data were diverse and conflicting as mentioned in a late nineteenth century review,[24] as the yield would have depended greatly upon the kind of tree, the method of harvesting the gutta percha, etc. In Table 1, the yield of gutta percha found in literature has been summarised. The value ranged about 0.2 – 15 kg per tree.

If the diameter, the length and the specific gravity of the trunk are assumed as 40 cm, 1500 cm and 0.75, respectively, then the yield of 2 kg and 5 kg, for instance, equals to 0.142 wt% and 0.354 wt%, respectively. Such magnitude of values is much lower than the 1.8 wt% of "crude gutta percha" obtained from the leaves of *Palaquium oblongifolium* by the modern sophisticated process at the Cipetir Factory. Note that the crude gutta percha is the mixture of pure gutta (*trans*-1,4-poly-isoprene) with resins, the composition of which will be detailed below.

Table 1 The yield of gutta reported in literature

Genus of tree	Year	Diameter (cm)*	Height (m)*	Yield (kg)*	Ref.
Tuban Tree	1847	-	-	12-3	25
Isonandra gutta, Hooker	1878	90	18	13-18	26
Getah Taban Merah	1885	60	30	1.055	27
Getah Taban Puteh,	"	60	-	9.84.	"
Getah Taban Simpor	"	51	19	5.44	"
Getah Sundik	"	75	-	17.5	"
Palaquium Treubii	1880s	20	-	0.24	28
"	"	13	-	0.16	"
"	"	70	-	0.19	"
"	"	-	-	1.40[†]	"

*) Unit converted into cgs from the British Imperial System.

[†]) V-shaped tapping.

The quantity of gutta percha exported from Singapore, the major port in the region, to Great Britain and the Continent from 1845 to 1847 amounted to 6,918 piculs, or 418.4 tons.[29] The importation of gutta percha into Singapore from 1885 to 1896 from producing countries[30] is shown in Figure 9. The total exportation amounted 27,507 tons for the twelve years. It means, if one assumes the yield of gutta percha as 5 kg per tree, as many as five and a half million trees were felled down in the forests of the Southeast Asian Area.[31]

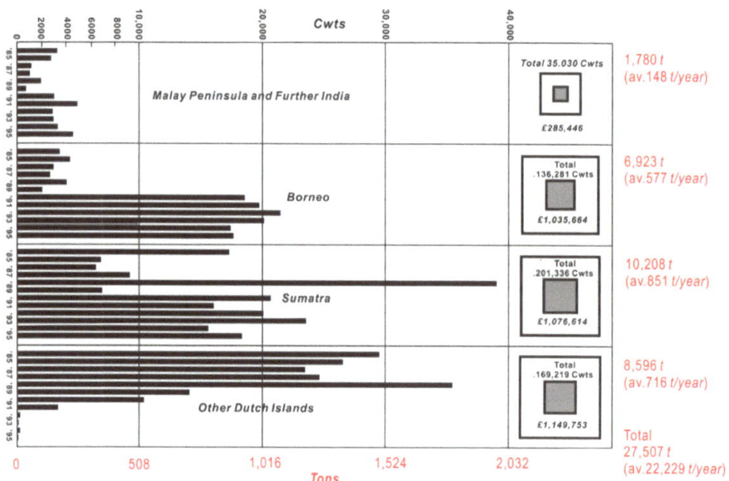

Figure 9 The importation of gutta percha into Singapore between 1885 and 1896 from producing countries.[32] The numerals in cgs unit below the bottom axis and at the right-hand side in red colour have been added by the present writer.

Chapter 2
Introduction to Europe; Processing and products

Introduction to Europe

In the catalogue, entitled, *Musaeum Tradescantianum: Or, a Collection of Rarities Preserved at South-Lambeth Neer London*, authored by John Tradescant and published in 1656,[33] is found an item, "The plyable Mazer wood, being warmed in water, will work to any form" (See Figure 10). From the year of publication, it is considered that the author, "John Tradescant" was John Tradescant the Younger (1608 - 1662), the son of John Tradescant the Elder (c.1570 - 1638), and that the person who brought the plyable Mazer Wood probably in the 1610s - 1620s from somewhere Asia was the father.

Figure 10 The catalogue of Musaeum Tradescantianum (1656). From the left, the cover, the portraits of John Tradescant the Elder and the Younger (father and son), the page listing the "plyable Mazer wood".[34]

Although it is almost certain that Mazer Wood was nothing other than gutta percha which was reintroduced to Europe nearly two centuries later, the Tradescant's Mazer Wood does not seem to have aroused much interest. It is probable that both the father and son of the Tradescants themselves might have thought it as a sort of special wood and not perceived its potential values as a useful new material, having classified it into the category of "Other Varieties of Rarities", not of "Seeds, Gummes, Roots, Woods, &c". People in those days did not think it important either. When W. Pinkerton, an Irish engineer, wrote an article, *Mazer Wood: Gutta Percha*, in 1851,[35] the specimen as well as Mazer Dishes, another item listed in the Tradescant's catalogue in the category of "Utensils and Householdstuffe", did no longer exist in the Ashmolean Museum at Oxford, probably having been judged valueless and abandoned.

In 1843, gutta percha was reintroduced into Europe independently by two persons. One was Dr. Jose d'Almeida, a doctor and naturalist who was a resident of Macau. In the April of that year, he brought with him a lump of solidified sap of a tree, indigenous in Singapore, which was said to "becomes ductile by being placed in hot water", as well as a riding whip made of the substance, and presented them to the Royal Asiatic Society, London. Although Dr. d'Almeida handed some portions of the raw material for analysis to two scientists, unfortunately neither of them seems to have done experiments on the substance.

The other person was Dr. William Montgomerie, a surgeon in Singapore. His specimens which he sent to Calcutta in 1842 were transferred to and arrived at the Society of Arts, London, in the summer of 1843. They consisted of "one bottle of the juice; specimens of thin sheets, resembling scraps of leather; specimens in a spongy mass as it concreted in a vessel; specimens of the substance formed into a mass by agglutinating the thin sheets by means of hot water". The Joint Committee of Chemistry, Colonies, and Trade of the society took Montgomerie's specimens into consideration and in a meeting in 1845, it was resolved "that this substance appears to be a very valuable article, and might be employed with great advantage in many of the arts and manufactures of the country" and the society awarded a Gold Medal to Dr. Montgomerie. Some criticisms that the honour should be shared by Dr. d'Almeida were ignored with various reasons or excuses.

The above story was written in detail in contemporary reviews.[36; 37] The portraits of Dr. d'Almeida and Portraits of Dr. Montgomerie are shown in Figure 11.

Figure 11 Dr. Jose d'Almeida Carvalho e Silva (left),[38] and Portraits of Dr. William Montgomerie (right).[39]

Processing and products

Let us now see how gutta percha was processed with reference to *Gutta percha, its discovery, history, and manifold uses*, published in 1851,[40] which included illustrations sketched at The Gutta Percha Company, jointly established in 1845 by Charles Hancock, brother of the great engineer Thomas, and Henry Bewley, a Dublin chemist.

The imported gutta percha was typically in the form of a block, shown in Figure 12 which was very tenacious mass. (In fact, the present writer experienced that a mass of gutta percha was not sectile, unable to be crushed by hammer, very hard to be cut by a knife, and that the best method to get small pieces for laboratory experiment was to saw the lump by a jigsaw.)

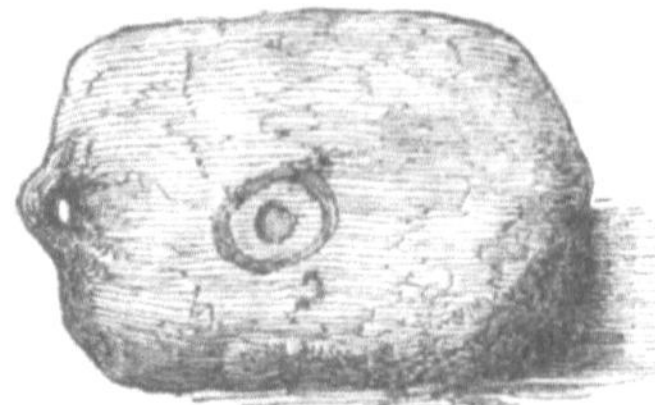

Figure 12 Block of gutta percha, a typical form of raw material, imported at that time (left) and a slice of it (right).[41]

First of all, the blocks are shaved into slices with the cutting machine (Figure 13 (1)), a large revolving iron disc equipped with sharp blades like a carpenter's plane.

To separate the pure material from the dirt and other extraneous material, slices are first boiled in a tank to soften and made into one mass, whereupon considerable impurities are left at the bottom. Next, the soft mass is thrown into what they called a "teaser" (Figure 13 (2)), a sort of large circular box, in which a cylinder, set all over with rows of bent jagged teeth, is revolving to tear the material to shreds. The shreds fall into a vat of water underneath. Here the Gutta Percha being lighter than water floats on the surface, while the impurities disengaged from it sink to the bottom.

The fragments of gutta percha, crisp and clean, are boiled again to turn into a warm soft mass and then passed to the "kneader", also called "masticators", which has a revolving drum with a cogged surface set in the cast-iron casing, kept hot by steam. This operation thoroughly expels every air bubble, solves every knotty point, and reduces the gutta percha to a uniform consistency.

The kneaded mass is carried to the "rolling machine" (Figure 13 (3)) which precisely resembles the one in a paper-mill. Gutta percha, placed on bands of felt, is passed between large steel cylinders, distant from each other by the thickness required; and during a long journey, up and down, over and under, it gradually cools and hardens, appears at the other end a smooth firm flat sheet, and is wound upon a drum until the requisite length has been worked off. The sheetings thus prepared are cut into appropriate sizes for stamping and moulding (Figure 13 (4)).

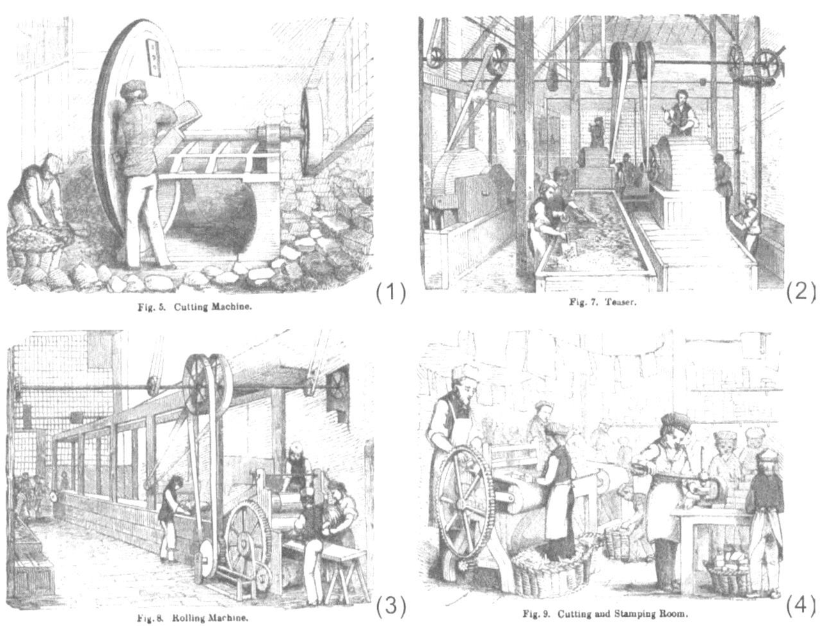

Figure 13 Processing machines of gutta percha. (1) Cutting machine, (2) Teaser, (3) Rolling machine, (4) Cutting and stamping room, in the book of 1851.[42]

The same book illustrated as many as 49 figures of gutta percha products, such as shown in Figure 14. Items of complicated shapes and engravings, Figure 15, were also manufactured in the later nineteenth century.

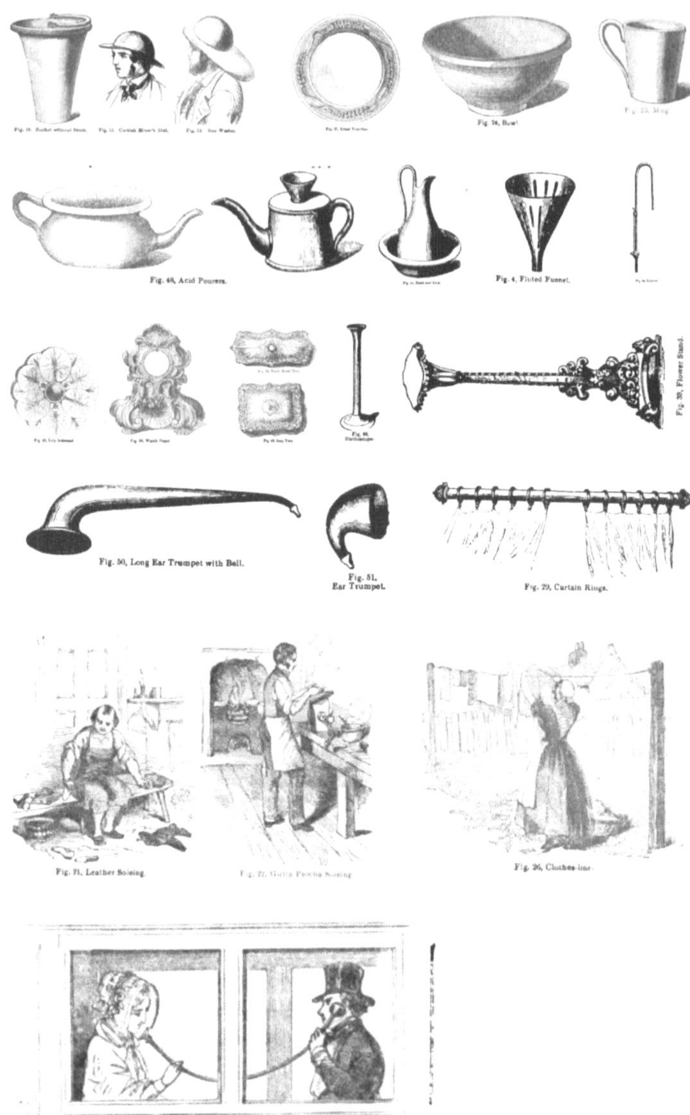

Figure 14 Various gutta percha products, extracted from the book of 1851.[43]

Figure 15 Gutta percha products of the Victorian Age. Buttons,[44] Unioncase with Napoleon portrait,[45] Hand mirror,[46] Chessmen.[47]

The Official Catalogue of the Great Exhibition of the Works of Industry of All Nations 1851, London[48] listed as many as twenty-nine companies and organizations which entered their products in various categories as summarised in Table 2 (The Catalogue itself was quite amazing to the present writer when it was downloaded from the internet and looked at. In the 300 page book, as many as 8,687 items entered from domestic companies and organizations as well as the Queen and His Royal Highness were classified into 30 categories, and those from thirty British dependencies and twenty five foreign states which were too many to be counted but presumably more than six thousand were well sorted. In fact, all these items were exactly placed in the booths set up inside the Crystal Palace, which was constructed in the Hyde Park, and outside. It is probable that the organisers had enormous brains which must have been as smart as modern computers).

Table 2 Participants with gutta percha items to the Great Exhibition of the Works of Industry of All Nations 1851, London. Summarised from the Official Catalogue.[49]

Class	Category	Items
Class 8	Naval architecture. Military engineering.	1
Class 8	Guns, weapons, &c	1
Class 9	Manufacturing machines and tools	1
Class 9	Agricultural and horticultural	1
Class 10	Philosophical, Musical, Horological, and Surgical instruments	6
Class 16	Leather, Saddlery, Boots and Shoes, Skins, Fur, and Hair	2
Class 17	Paper, Printing, and Bookbinding	1
Class 19	Tapestry, Floor clothes, Lace, and Embroidery	1
Class 22	General hardware, including Locks and Grates	3
Class 25	China, Porcelain, Earthenware. &c.	1
Class 26	Furniture, Upholstery, Paper hangings	1
Class 28	Manufactures from animal, and vegetable substances	6
Class 29	Miscellaneous manufactures, and small wares	1
Abroad	India	1
"	Canada. Nova Scotia. Newfoundland	1
"	New Sealand Labuan	1
	Total	29

At The International Exhibition of 1862, London, The Gutta Percha Company presented a great many products, as listed in Figure 16.[50]

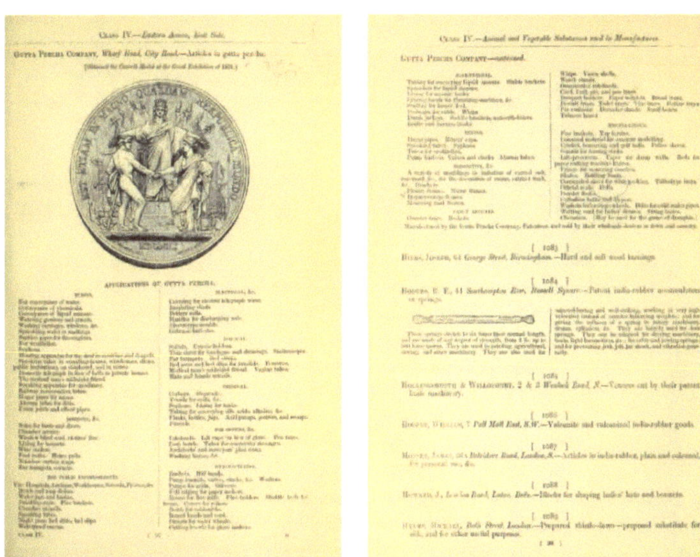

Figure 16 The page of The Gutta Percha Company in the Illustrated Catalogue, The International Exhibition of 1862, London, with the Council Medal awarded at the 1851 Exhibition.[51] On the medal is engraved EST ETIAM IN MAGNO.

Details of some processing machines are shown in Figures 17, 18, 19 and 20.

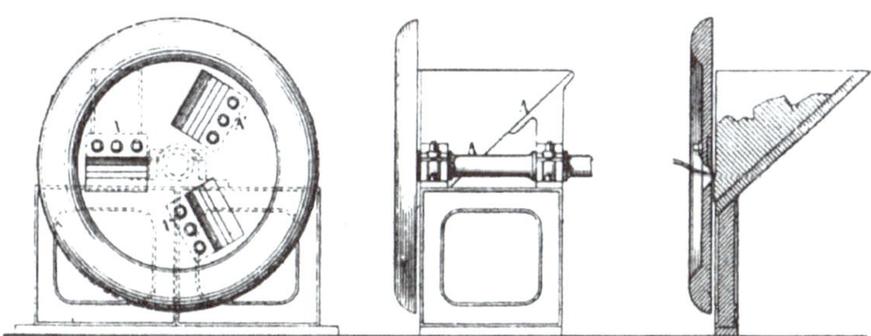

Figure 17 Cutting machine to slice the block of raw gutta percha (Hancock's system).[52]

Figure 18 A type of Masticator (Truman's masticator) and an illustration to show the principle.[53]

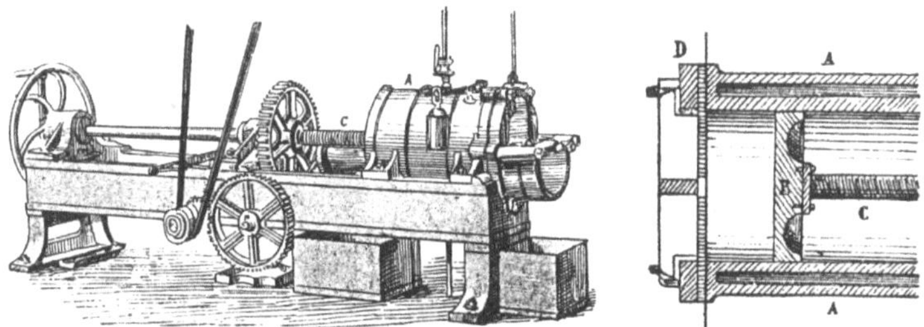

Figure 19 Horizontal filter press (strainer) to ensure the dryness.[54]

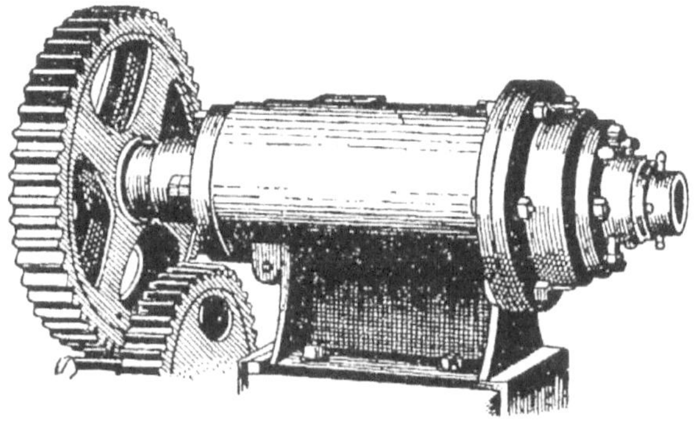

Figure 20 Machine by Friedrich Krupp Aktiengesellschaft Grusonwerk, for making tubing by extrusion. The strong steel barrel contains a revolving spiral, which forces the rubber through a shaped die or nozzle.[55]

It ought to be remarked that various plastic processing machines which are of use today are based on the principles contrived in the nineteenth century for gutta percha.

The gutta percha "purified" by the physical method mentioned above would have been not pure gutta (C_5H_8, *trans*-1,4-polyisoprene) but a mixture with two other substances, alban ($C_{20}H_{32}O_2$) and fluvial ($C_{20}H_{32}O$), which were contained in the latex, the average composition being:

Gutta   78 to 82 %,

Alban   16 to 14 %,

Fluvial   6 to 4 %.[56]

Alban and fluvial can be dissolved away by boiling in alcohol. The former which precipitates upon cooling can be collected by filtration and the latter, obtained by evaporating the remaining alcoholic liquid.[57]

With regard to the purification of commercial gutta percha, the following method was given in literature.[58]

One part gutta-percha is put in a bottle containing sixteen part of carbon disulphide and the bottle is frequently shaken until dissolved. A small quantity of animal charcoal is added to and shaken, then the solution is filtered. In four part of alcohol contained in a wide-mouth vessel, the filtrate is poured, a little at a time, and shaken each time to precipitate gutta. At the end if the liquid phase is turbid, add more alcohol and shake until the liquid become transparent. Place the precipitated gutta on a pill-plate and press it with a spatula into a sheet as thin as possible, and leave the sheet for 24 hours or until all smell of carbon disulphide and alcohol has disappeared. Carbon disulphide and alcohol can be recovered by distillation.

As the solvent, chloroform was also used[59]. Pure gutta was not only more solid and homogeneous but also tougher than the raw gutta percha.[60]

The chemical purification expanded the application to the dental area, viz. denture and dental fillers as depicted in Figure 21.

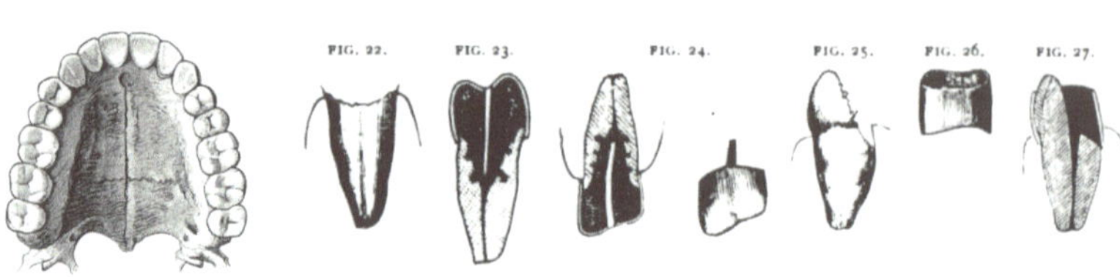

Figure 21 Dental applications.[61]

Gutta percha, the only chemical-resistant natural resin, was exclusively used for gaskets and vessels. Until the mid-1970s when polypropylene bottles became common, gutta percha bottles of hydrofluoric acid (Figure 22) were seen in every chemical factories and school laboratories. For the present writer and his contemporaries, it was common to use hydrofluoric acid for marking glass pieces which they made by glass-blowing: To do this, a part of the glass surface is coated with wax, marks are inscribed by means of an iron pen, the part is covered with a piece of paper, a few droplets of the acid is applied to etch the marked part, after a while the glass piece is washed and finally the wax is removed with a solvent.

Figure 22 Gutta percha-made hydrofluoric acid bottle.[62]

The use of gutta percha for golf-balls was remarkable. Without gutta percha, golf would not have become a popular sport as was reviewed elsewhere. A chronological summary found in an article of The Times[63] has been duplicated in Figure 23.

History of the golf ball

Before the 17th century
Possibly solid hardwood

17th to early 19th centuries
The "featherie", made of goose feathers stitched up inside a leather sphere

1848 to 1900 The "gutty", a solid ball of gutta-percha resin obtained through boiling the sap of a tropical tree

1899 to 1960s The Haskell, with a core of tightly wound rubber in an outer shell of balata resin or a thermoplastic called Surlyn. Moulded into a dimpled pattern so that it would roll better than the previous ball

1970s to present
Surlyn case around a heart of polybutadiene, a soft, bouncy, petroleum-based polymer

Figure 23 The history of the golf ball (The Times, November 25 2017).[64]

The line of "1899 to 1960s" stated that "balata" was used for the shell of ball but gutta percha was actually better. Balata, a lower molecular-weight *trans*-1,4-poly-isoprene, is a cheap substitute obtained from *Sapotaceae Manilkara bidentata*, indigenous to South America (cf. Chapter 6).

Chapter 3
Insulative covering of submarine cables

According to literature,[65] the insulating property of gutta percha was first observed in 1846 by Werner von Siemens (1819 - 1892). Michael Faraday (1791–1867) also noticed this property and in 1848 published a letter entitled: "On the Use of Gutta Percha in Electrical Insulation",[66] which was addressed to Richard Phillips: He wrote, "I have lately found gutta percha very useful in electrical experiments... A good piece of gutta percha will insulate as well as an equal piece of shellac, whether it be in the form of sheet, or rod, or filament; but being tough and flexible when cold, as well as soft when hot, it will serve better than shellac in many cases where the brittleness of the latter is an inconvenience ...."

Faraday also mentioned that not all gutta percha obtained from the manufacturer's hand was not equally good but it was brought into the best state if a portion was warmed in a current of hot air and be stretched, doubled up, and kneaded for some time between the finger to dissipate the moisture within. Otherwise, it was submitted to a gradually increasing temperature, at about 350° or 380°F (177° or 193°C).

For the seamless covering of wires with gutta percha, Siemens designed a small screw-press illustrated in Figure 24 and used it in 1848 for the manufacture of cables for a subterranean telegraph line of 2.5 miles (4.0 km) long, from Berlin to Gross Beeren.

Figure 24 Gutta-percha press for the seamless insulation of electrical conductors jointly developed by Werner von Siemens and Johann Georg Halske, 1847. Siemens Corporate Archives, Munich, Germany.[67]

Figure 25 illustrates the principle of the machinery used for covering wires with gutta percha, in which the gutta percha contained in the space between the piston and the bottom of the cylinder was forced through a die, *d*, where it surrounds the wire, which then enters a long trough with cold water, in which it passes to and fro several times, until the gutta percha became sufficiently cool and hard.

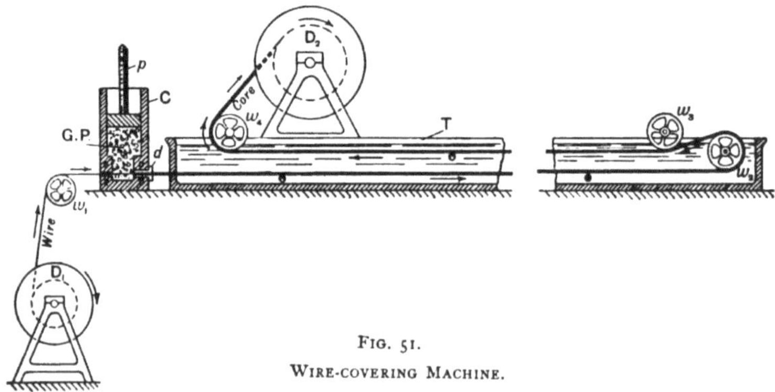

Figure 25 A wire-covering machine designed by Werner Siemens.[68]

Figure 26 shows an advanced cable covering process which is equipped with a screw extruder for supplying softened gutta percha.[69]

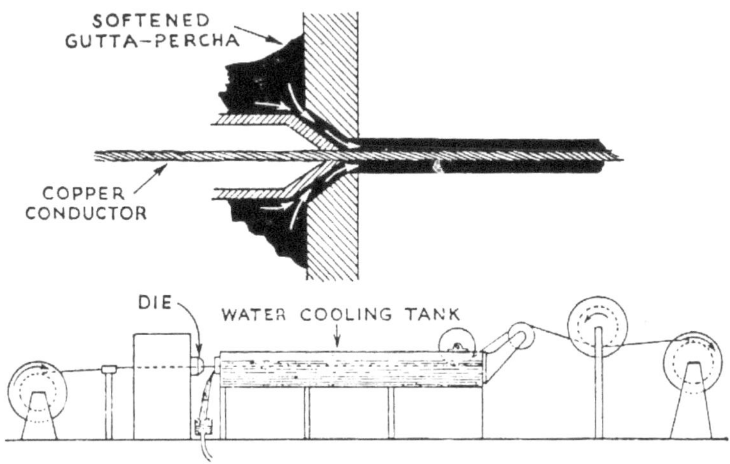

Figure 26 Covering electric cables with rubber or gutta-percha insulation by the extrusion process.[70]

In 1851, the first submarine cable was successfully laid across the English Channel between Dover and Calais, after the unsuccessful attempt in the previous year. In the Netherlands East Indies, the country where most gutta percha came from, it was as early as 1859 that Java and peripheral islands were linked with submarine cables.[71] In Japan, it was 1872 that submarine cable was laid for the first time between Shimonoseki (westend of Honshu) and Moji (northeast tip of Kyushu).

In 1866 the transatlantic cable connection was completed by employing SS Great Eastern (18,915 tons), the greatest ship at that time, designed by Isambard Kingdom Brunel. The cable was manufactured by WT Henley's Telegraph Works Company Limited, Greenwich, London, with the core made by Telegraph Construction and Maintenance Company.[72] In the Pacific, the links from San Francisco to Manila, Shanghai and Tokyo, via Hawaii and Guam, were established in 1903, 1906 and 1906, respectively, to complete the world-wide telegraph cable network.

Figure 27 and Figure 28 show the structure of gutta-percha insulated cables.

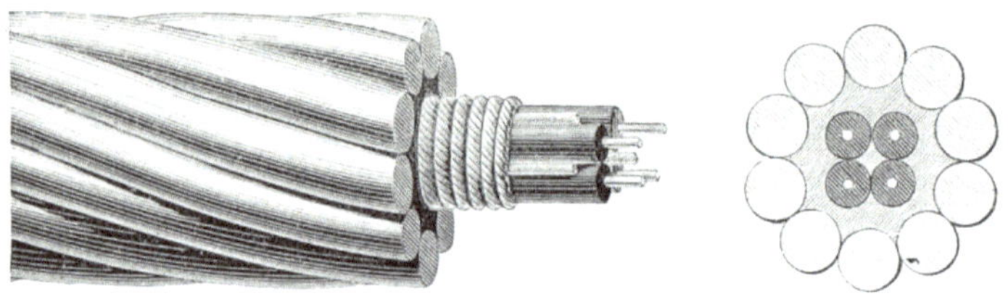

Figure 27 The Dover-Calais Submarine Cable, 1851, with double gutta-percha layers, manufactured by The Gutta Percha Company.[73]

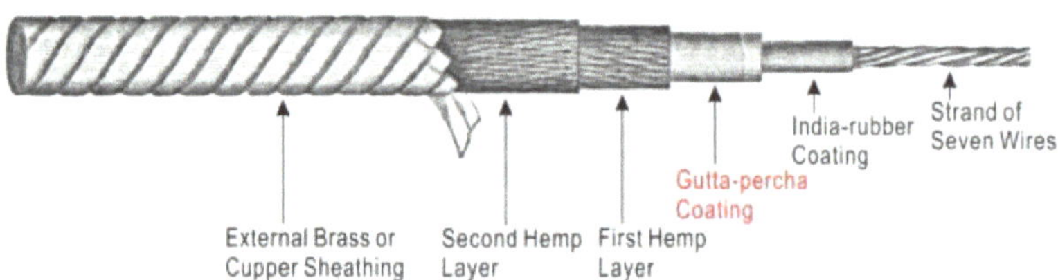

Figure 28 Siemens's Submarine Cable (probably the type used for the transatlantic cable in the 1870s).[74]

Figure 29 illustrates a scene of cable manufacturing at the Telegraph Construction and Maintenance Company, Greenwich, ca. 1865.

Figure 29 Cable manufacturing with gutta-percha at the Telegraph Construction and Maintenance Company in Greenwich, London, ca. 1865.[75]

The material used for the cable insulation in those days was allegedly raw gutta percha, as it was written that "the electrical properties, i.e. the insulation resistance and inductive capacity, are very little affected by the extraction of the resin, unlike the physical and mechanical properties".[76] Nevertheless, pure gutta such that produced in the later decades at the modern factory at Cipetir, West Java, must have been much better for the processing as well as the uniformity of the products.

Figure 30 shows the premises of The India Rubber, Gutta Percha and Telegraph Works Company, Silvertown, Newham, London, ca. 1869, in which The Silvertown Telegraph Cable Works is seen in the right-hand side.[77] The picture gives an impression that the rubber/gutta percha industry was extremely prosperous in those days.

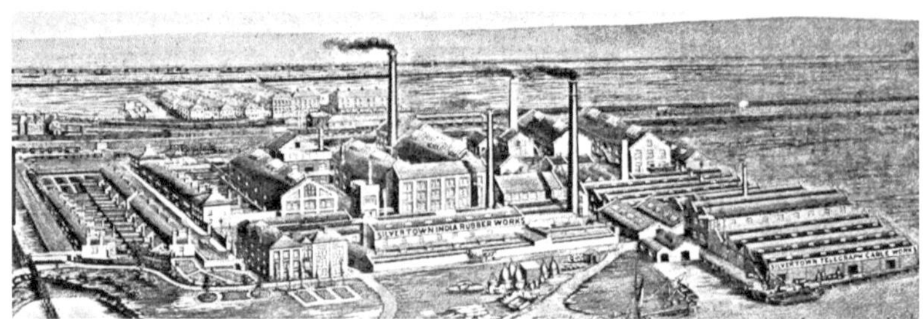

Figure 30 A view of The Silvertown Rubber Works (left to middle) and Silvertown Telegraph Cable Works (right). Ca. 1869.[78]

Chapter 4
Structure and properties

In this section some works on the structure and properties of gutta percha are arbitrarily reviewed.

As aforementioned, in 1860 Greville Williams considered that the results of the action of heat upon caoutchouc and gutta percha to be merely the disruption of a *polymeric body* into substances (isoprene) having a simple relation to the parent hydrocarbon.[79] By 1870s, raw gutta percha was determined to be the mixture of gutta (C_5H_8, 78-82%), alban ($C_{20}H_{32}O$, 16-14%), fluvial ($C_{20}H_{32}O_2$, 6-4%).[80] It was confirmed that the physical properties depends much on the purity. Pure gutta becomes pliable at 25 °C and gradually softens on heating: At 60 °C, it is so plastic that it may be pressed into any shape or drawn into threads. The specific gravity was determined as 1.01 - 1.02.

In 1942, C. W. Bunn analysed the structure of natural rubber and gutta percha (Figure 31) by X-ray diffraction and determined the conformation on planar models as shown in Figure 32.[81]

$$\left(\begin{array}{cc} CH_3 & H \\ \diagdown & \diagup \\ C=C \\ \diagup & \diagdown \\ -CH_2 & CH_2- \end{array} \right)_n \qquad \left(\begin{array}{cc} CH_3 & CH_2- \\ \diagdown & \diagup \\ C=C \\ \diagup & \diagdown \\ -CH_2 & H \end{array} \right)_n$$

Cis-Poly-isoprene   *Trans*-poly-isoprene

Figure 31 Structural formula of natural rubber and gutta percha, 1,4-*cis* and 1,4-*trans*-poly-isoprene.

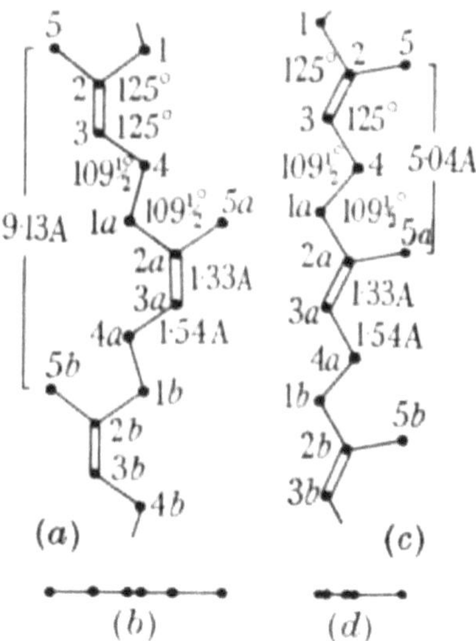

Figure 32 Planar poly-isoprene models by C. W. Bunn. Left: *Cis*-chain, Right: *Trans*-chain.[82]

In 1938, K. H. Storks[83] published a theory on "the folding habit of linear polymers", such as polyethylene succinate, polyethylene adipate and gutta percha, basing on the electron diffraction, almost two decades earlier than that raised by A. Keller[84], P. H. Till Jr.[85] and E. W. Fischer[86] with linear polyethylene obtained with a Ziegler Catalyst.

Storks wrote, "It is surprising that most of the crystallites are oriented with their fibre axis directions normal to the plane of a film the thickness of which is much less than the length of a macromolecule...The gutta-percha molecule may possibly fold by a mechanism of rotation around the single bonds. The chemical repeating units of this polymer is short and relatively few folds per macromolecule are required in a film of 200Å thick..." (Figure 33)

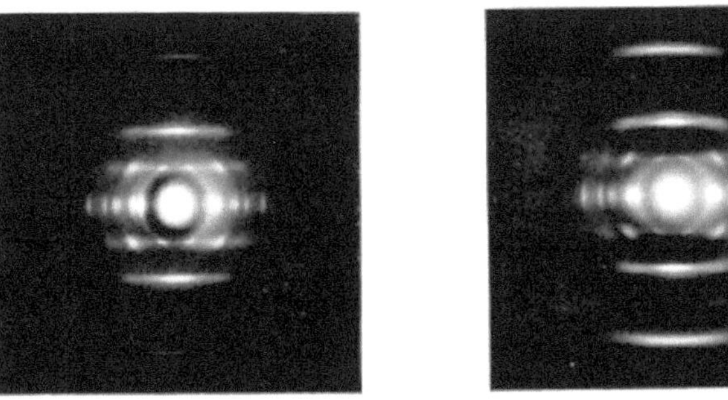

Fig. 4 —Electron diffraction pattern from stretched
"Br-4" grade gutta-percha.
$(L\lambda = 2.73_{(0)} \times 10^{-6}$ mm.^{2}.)

Fig. 5 —Electron diffraction pattern from stretched
"white" gutta-percha.
$(L\lambda = 2.70_{(0)} \times 10^{-6}$ mm.^{2}.)

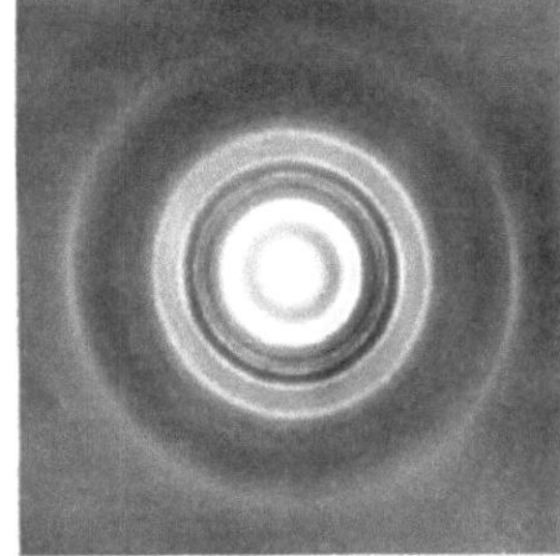

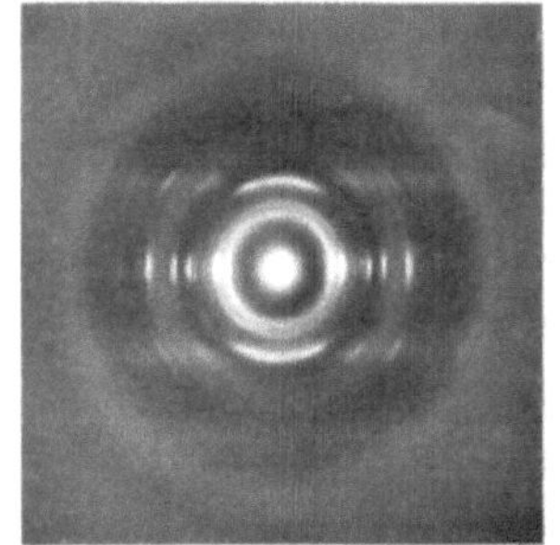

Fig. 8 —Electron diffraction pattern from unstretched
"white" gutta-percha at normal incidence
$(L\lambda = 2.70_{(0)} \times 10^{-6}$ mm.^{2}.)

Fig. 9 —Electron diffraction pattern from unstretched
"white" gutta percha at 45° incidence
$(L\lambda = 2.70_{(0)} \times 10^{-6}$ mm.^{2}.)

Figure 33 Electron diffraction examination of linear high
polymers by K. H. Storks (extract).[87]

In the recent decades, detailed crystal structure as well as the
elastic properties was determined by Tashiro et al. as shown in
Table 3 and Figure 34.

Table 3 Crystal structure and elastic modulus of natural rubber and gutta percha.[88]

		Rubber	Gutta percha	
			α-form	β-form
Crystal form		Mono*	Mono*	Ortho†
Unitcell parameter	a	12.1566	8.0412	7.5901
	b	8.939	5.8656	11.7425
	c	8.0041	8.8461	4.7036
	α	90	90	90
	β	879923	104.1431	90
	γ	90	90	90
	ν	869.2552	404.5922	419.216
Monomer units/cell		8	4	4
Density		1.041	1.1183	1.0793
Elastic modulus	bulk	8.9634	8.8925	9.0074
	x	8.8109	8.3734	8.86289
	y	8.857	8.232	9.0085
	z	121.8575	118.5674	120.6373

*) Mono=Monoclinic, †)=Orthorhombic, ν =Volume

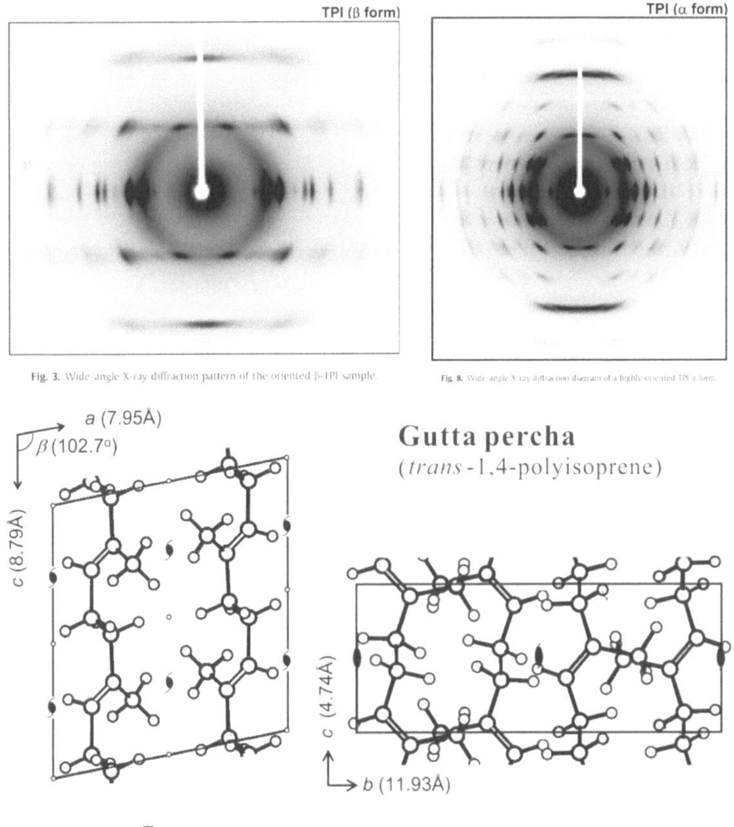

Figure 34 Diffraction patterns of α- and β-form *trans*-isoprene (gutta percha) and their crystal structure. Results of X-ray diffraction.[89]

The temperature dependence of endotherm by DSC, X-ray diffraction and infrared absorption was measured by Paramita J. Ratri and K. Tashiro (Figure 35).[90]

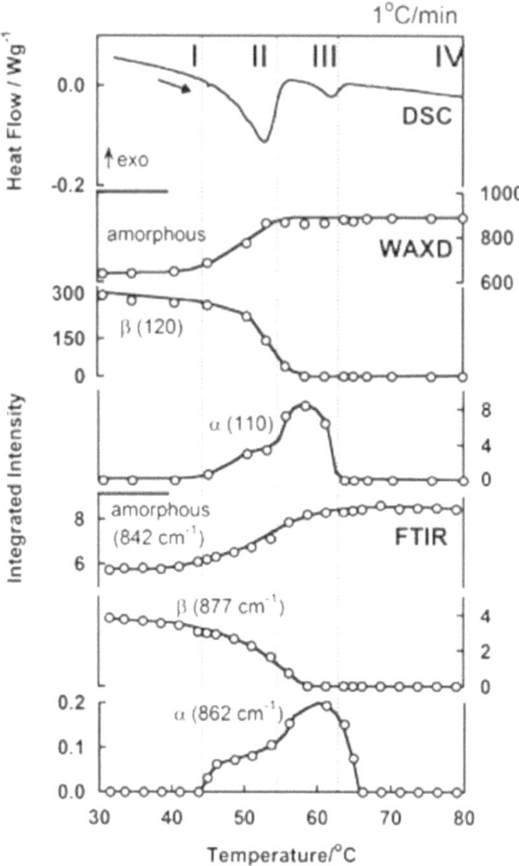

Figure 35 Temperature dependence of endotherm by DSC, X-ray diffraction intensity and infrared absorption intensity.[91]

GPC measurement was conducted at the Prof. P. J. Lemstra's laboratory in Eindhoven University of Technology in 2005 for white gutta from Cipetir. The curve showed a smooth maximum from which the M_w = ca. 3.5 x 10⁵ was obtained but the original chart was unfortunately lost. GPC measurement was repeated in 2018 by courtesy of Prof. Toshiaki Ougizawa, Tokyo Institute of Technology, and a curve with a maximum which gave almost the same M_w value was reproduced as shown in Figure 36.

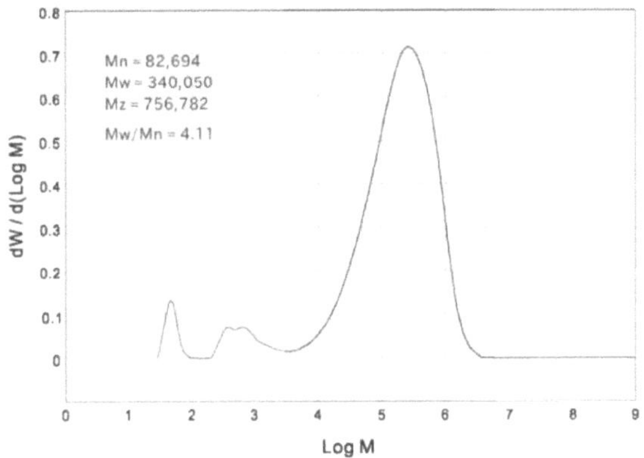

Figure 36 GPC trace of white gutta from Cipetir. Mw=3.4 x 10^5 in the molecular weight range: $3.0x10^3 – 6.6x10^6$. By courtesy of Prof. Toshiaki Ougizawa (2018).

Recently, Hanna Szczepanowska and Velson Horie reported on the Raman spectroscopy of gutta percha, natural rubber and balata, which was conducted aiming at establishing a method to identify cultural heritage artefacts. As to P. gutta, it was shown that strong peaks observed with fresh latex became weak when coagulated and the bands were very weak in a 40-year old specimen,[92] presumably as a result of oxidative degradation (the present writer's conjecture). See Figure 37.

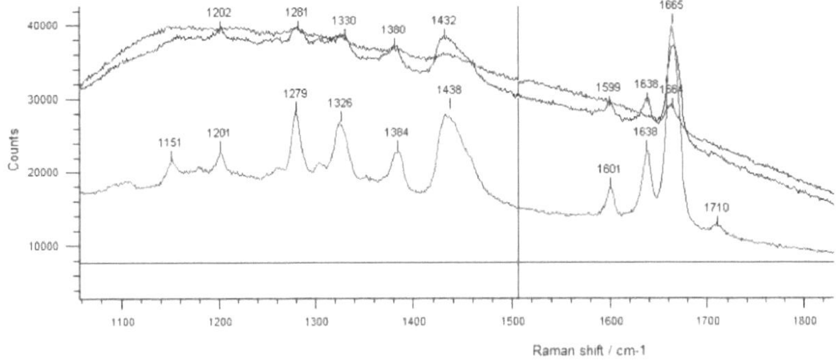

Figure 37 Raman spectra of P. gutta. Bottom: Fresh latex, Middle: Coagulated, Top: A 40-year old sample.[93]

Chapter 5
The founding of plantation at Tjipetir

With the increased demand for gutta percha, trees in many forests in the island of Singapore and Malay Peninsula rapidly diminished. Attempts to culture the trees, done in Singapore as early as in 1848 and subsequent years were not successful.[94; 95] The family of the plant was mainly of *Isonandra*. Then, gutta percha-producing plants were widely surveyed. Figure 38 shows the distributional pattern of the genus Palaquium that extends from India and Ceylon to the southwest Pacific area.

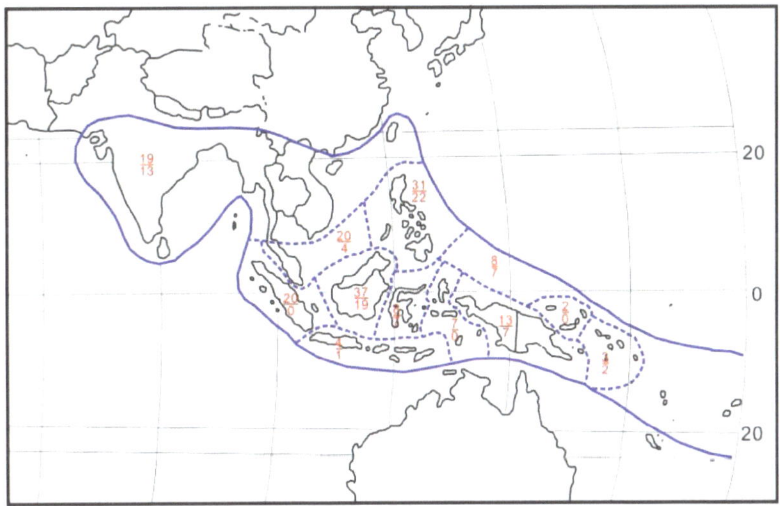

Figure 38 The distribution of genus Palaquium. Numerals above bar indicate number of species, below bar number of endemics.[96] Redrawn by the present writer on CorelDRAW.

In Java, the Buitenzorg Botanical Garden received young *Isonandras* in 1847. In 1858, Mr. J. E. Teijsmann, the famous gardener, obtained 2,000 young trees from West Borneo and distributed them to three places in Java. Most of them died but 80 of them planted at Purwokerto reached maturity.[97] In 1884, Dr. William Burck, under the eminent director Dr. Melchior Treub, collected *Palaquium oblongifolium* and several other gutta percha-producing plants (*Palaquium gutta, Palaquium Treubii, Palaquium Borneense, and Payena Leera*)[98] and the new gutta percha-producing plants were planted in the Garden, as highlighted in the guidebook of the garden authored by Dr. Burck himself.[99] A *Palaquium oblongifolium*, Burck tree still survives in the Bogor Botanical Garden (Figure 39).

Figure 39 *Palaquium oblongifolium* Burck in the Bogor Botanical Garden planted in the 1880s (Photographed by the present writer, June 2019).

In 1885, a vast Experimental Garden was opened by the Dutch Government at Tjipetir, Preangan Regency (the present-day Cipetir, Kabupaten Sukabumi), at an elevation of about 1,300 feet above the sea-level, and extending over 250 acres, and a large-scale research project, started. Among various genera, *"Palaquium oblongifolium*, Burck"*, which was determined to be the best of all, bore abundant fruit in 1895.[100] The grafting which was said to be impossible in the past was successfully done.

The method to obtain high-quality gutta percha from harvested leaves and twiglets (called leaf-gutta or white gutta) was established and a large factory equipped with modern machines and chemical apparatus, constructed, as will be described in the next section.

According to literature,[101] the first lot of gutta-percha, from plantation trees, amounting to 13,520 kg (29,736 lbs), was produced in 1912 at the Netherlands East India Government estate of Tjipetir in Java. During that year also, 23,500 kg (51,707 lbs) of plantation-gutta was also produced in Sumatra, in addition to minor amounts in Borneo and some of the small neighbouring islands. From 1912 to 1924, the annual production of leaf gutta in the entire Netherlands East Indies fluctuated between 25,910 kg (57,073 lbs) and 326,050 kg (717,317 lbs). In 1948, exports of gutta from Java (thought to be mainly from the Tjipetir plantation) to the United States and Britain, amounted to approximately 131,931 kg (290,250 lbs). The present writer wished to obtain more detailed statistics and enquired the Indonesian authority which inherited the plantation but no old records remained. With regard to the description in the above citation that "plantation-gutta was also produced in Sumatra, in addition to minor amounts in Borneo and some of the small neighbouring islands", no information was available either.

After 1960s, the production of gutta-percha suffered from the emergence of polythene and other synthetic hydrocarbon polymers.

Chapter 6
Other sources of *trans*-1,4-poly-isoprene

(1) Balata and Chicle

Among Sapotaceae family, two more plants which generate *trans*-1,4-polyisoprene are known. One is Manilkara bidentata, native to northern South America, Central America and Caribbean islands.[102] The other is Manilkara chicle of Mexican and Central American origin.[103] Figure 40 shows their botanical illustration.

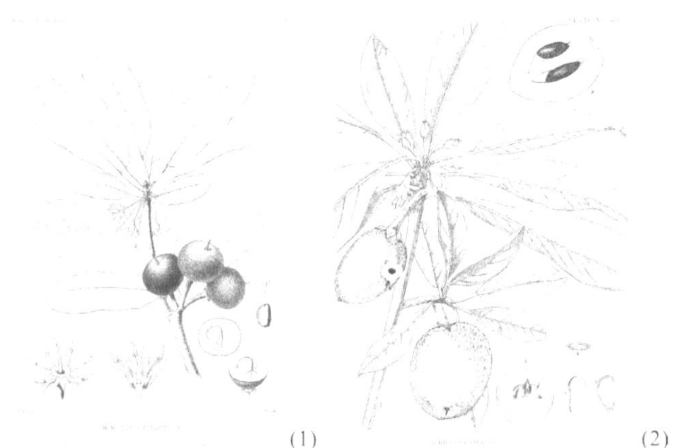

Figure 40 Botanical illustration of *Manilkara bidentata* balata (1),[104] and *Manilkara Zapota* chicle (2).[105]

The usefulness of balata was discovered in Dutch Guiana in 1856 and the first commercial product from British Guyana was exhibited at the London International Exhibition in 1862. A huge quantity was produced in all Guinea, Venezuela and other areas in subsequent decades, at the sacrifice of trees in the jungle. The balata tree has a straight trunk, the height and diameter of which attains 22.5 - 37.5 m (75 – 125 feet) and 1.275 m (4.25 feet), respectively. The latex contained in cells in the bark could be collected by tapping the trunk because it is thin and fluid, unlike in the case of gutta percha, but felling of trees was also conducted.

The coagulated material contains about 40% *trans*-1,4-poly-isoprene and about 40% resin, besides moisture and impurities.[106] The *trans*-1,4-poly-isoprene of balata was not useful as "thermoplastics", because the molecular weight was low (e.g., 51,000[107]). Thus, it was applied only in the form of vulcanisates to belting, shoe soles, golf-balls, etc.

Chicle is obtained by the tapping method. The material consists of about 17.4% hydrocarbon, 40% acetone soluble resin and 35% occluded water.[108] The hydrocarbon is a mixture of cis- and *trans*-1,4-poly-isoprene, 25% and 75%, respectively.[109] Although originally introduced as natural rubber and gutta percha substitutes, deresinated chicle has become important as the base for chewing gum.

(2) Tu-chung (Eucommia ulmoides)

According to The Yearbook of 2008 of International Dendrology Society,[110] this plant of central China origin was sent to Kew sometime between 1887 and 1890 by Dr. Augustine Henry who was stationed in China as a medical officer. Dr. Daniel Oliver described the feature of 杜仲, *Eucommia ulmoides* in 1890 as follows.

'The most singular feature about the plant is the extraordinary abundance of an elastic gum in all the younger tissues – excepting perhaps the wood proper – in the bark (in the usual sense of the word), the leaves and petioles, and pericarp; any of these snapped across, and the parts drawn asunder, exhibit the silvery sheen of innumerable threads of this gum... The bark, Dr. Henry, under No. 3182 wrote, 'is a most valued medicine with the Chinese, selling at 4s. to 8s per lb'. Under No. 4683 (the cultivated Patung specimens), he says further: 'It is planted from the seeds (fruit). The tree is cut down in the third to sixth Chinese months and stripped of its bark ...' "

Figure 41 shows the fine silk-like thread of gum in a broken leaf of Eucommia ulmoides and Figure 42, images of the same tree and its female and male flowers.

Figure 41 A leaf of *Eucommia ulmoides* broken to show the fine silk-like thread of gum.[111]

Figure 42 *Eucommia ulmoides* (1930-91101) at the Kew Garden in 2008.[112] A wide view, the female flowers and male flowers.

Major John Parkin wrote in 1921, "Eucommia, however, is exceptional in having its gutta-like substance existing in the dry solid state in the living plant; so that no milky juice exudes from any part when punctured."

In the same year, the gum of *Eucommia ulmoides*, extracted in 1919 from the twigs of a tree grown at Kew Gardens, was presented for The Gutta Percha Company for investigation. The amount of gum obtained was only about 1%.

In recent years, *trans*-1,4-poly-isoprene from Tu-chung has been extensively studied by a group of Hitachi Zosen Corp., in cooperation with Osaka University. The content of the polymer in various parts of the plant was investigated and, as seen in Table 4, it was revealed that the pericarps and seeds are most rich of the polymer, containing 13-20 and 25-32 %, respectively.

Table 4 The content of *trans*-1,4-poly-isoprene in various parts of *Eucommia ulmoides*.[113]

Organs	Leaves	Barks	Roots	Seeds	Pericarps
Content, %	0.6-2	12-8	10-5	13-20	25-32

Figure 43 shows the flowers and leaves of Tu-chung, and the silky threads drawn out of a leaf.

Figure 43 Flowers and leaves of *Eucommia ulmoides*, and the silky threads drawn out of a leaf.[114]

The molecular weight of *trans*-1,4-poly-isoprene from Tu-chung was found as high as $M_w > 2 \times 10^6$ and the substance was tougher than gutta-percha, the molecular weight of which was $M_w =$ ca. 3.4×10^5, although other properties were more or less similar. Making use of this property, the joint team of Hitachi Zosen and Osaka University developed artificial bones with the Tu-chung *trans*-1,4-poly-isoprene. The fact that the specific gravity of which, ca. 0.95, is smaller than ceramics and titanium is said advantageous.[115]

(3) Synthetic trans-1,4-poly-isoprene

After the discovery of the Ziegler- or Ziegler-Natta catalysis, the polymerisation of isoprene as well as other dienes and olefins were rigorously studied by Natta and his co-workers, Philips Petroleum and other groups. To the present writer's knowledge, synthetic *trans*-1,4-poly-isoprene is currently manufactured and marketed only by Kuraray Co., Japan, which started research in the 1970s.[116; 117] The molecular weight distribution is rather narrow and the average molecular weight is ca. 2.5×10^5. Other properties are similar to those of gutta percha. The polymer is said to have been applied to various areas, including dentistry, which have traditionally been covered by gutta percha, but the scale of the production is not disclosed.

Chapter 7
A guide to the gutta percha plantation/factory at Cipetir

The gutta percha plantation/factory at Cipetir has been maintained in operational conditions up until today by the efforts of PT Perkebunan Nusantara VIII (The No. XIII Plantation Corporation), Indonesia. Let us see the recent situation with photographs, the most of which had been taken by the present writer in 2002 - 2007.

Figure 44 and Figure 45 show a wide view of the factory and a part of Palaquium oblongifolium forest.

Figure 44 Cipetir Factory. A wide view.

Figure 45 A part of the Palaquium oblongifolium forest.

Nursery

Figure 46 and Figure 47 show the seeds and kernels of *Palaquium oblongifolium*, Burck.

Figure 46 Heap of the seeds of *Palaquium oblongifolium*, Burck, by courtesy of Prof, Malcolm R. Mackley who participated in the International Workshop on Green Polymers 1996.

Figure 47 Seeds and kernels of *Palaquium oblongifolium*, Burck. Diameter ca. 1.5-2 cm.

Figures 48 and 49 show the seedlings, and the grafted young plants which are ready to be planted in the forest.

Figure 48 Seedlings of *Palaquium oblongi-folium*. A wide view (left) and a close-up (right).

Figure 49 The grafted plants of *Palaquium oblongifolium*.

The young seedlings are used as the stock for grafting scion. In this nursery, not only *Palaquium oblongifolium,* Burck, but also other gutta-percha producing species, as well as Balata (*Manilkara bidentata*), a plant of South-American origin which produces lower molecular weight *trans*-1,4-poly-isoprene, are preserved.

Mature trees are pruned to limit the height and let branches grow in horizontal direction for the convenience of harvesting the leaves together with twiglets, such as shown in Figure 50.

Figure 50 A matured *Palaquium oblongifolium* tree.

Factory

The factory consists of Mechanical (Physical) and Chemical Divisions. The whole process is schematically shown in Figure 51.

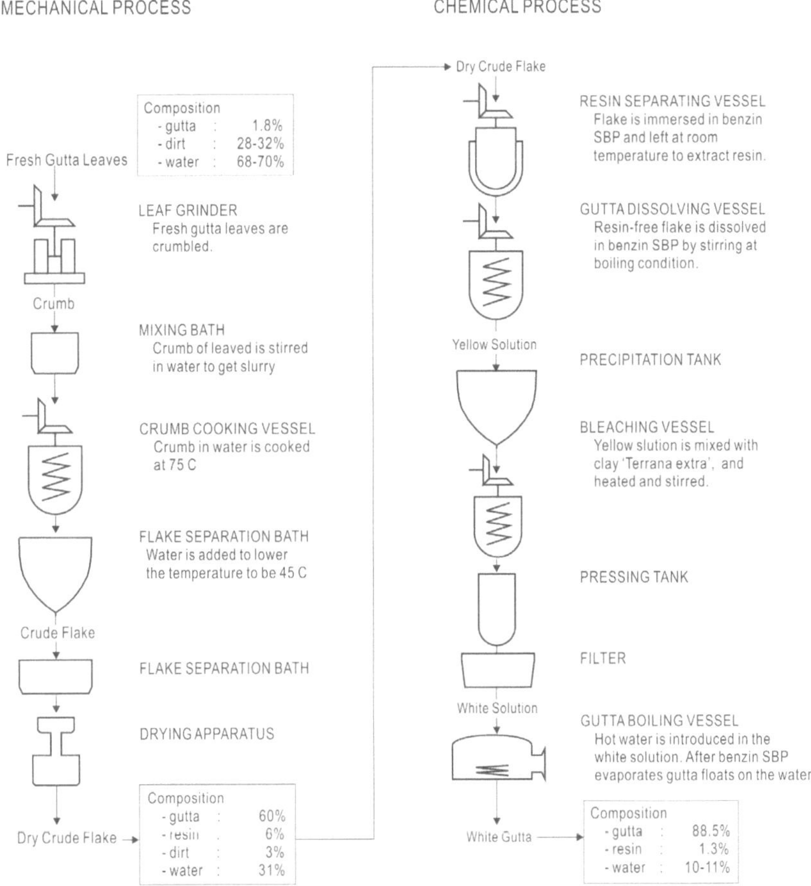

Figure 51 Schematic drawing of the chemical process (from the plantation's brochure, redrawn and the legends translated form Indonesian).

Mechanical Division (Physical process)

Fresh leaves are crumbled by means of large vertical roller mills, as seen in Figure 52 and Figure 53.

Figure 52 Heap of harvested leaves.

Figure 53 Vertical roller mills (left) and crumbled leaves.(right).

The heavy instruments are said to have been carried up to the plantation site of 400 m elevation from the sea-level from the Tandjoeng Priok Port of Batavia (Jakarta) by oxen-pulled carriages.

Crumbling is important for destroying the latex tubes and the surrounding cells to facilitate the exuding of the latex.

Crumbled leaves are put in the water bath of 75 °C and stirred. Crude gutta percha exudes out of the crumb and coagulates. The crude gutta percha which floats on the water surface in the cooling bath is scooped up and crumbled. See, Figures 54, 55 and 56.

Figure 54 Crum cooking baths.

Figure 55 Crude gutta percha floating on the water surface in the cooling bath (left) and the appearance of scooped-up gutta percha (right).

Figure 56 Grinders (left) and crumbled crude gutta percha(right).

Chemical Division (Chemical process)

Figure 57 shows a view of the chemical plant.

Figure 57 The chemical plant.

The crude gutta percha is immersed in cold benzine (petroleum ether) to extract resins in it and filtered. Figure 58 shows the resin-free gutta percha sampled from the line.

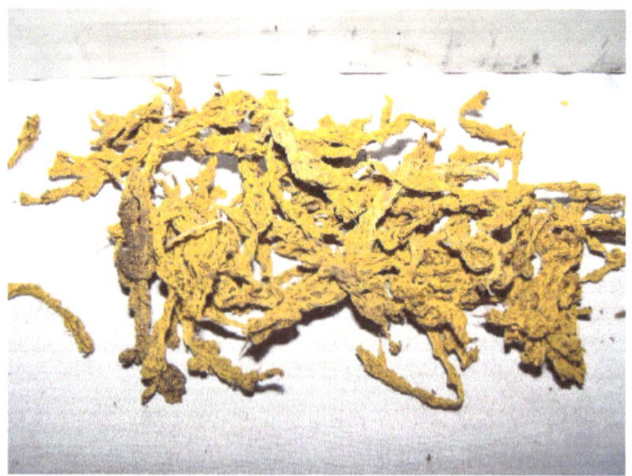

Figure 58 Resin-free gutta (sampled from the line).

In the line, the resin-free gutta percha is dissolved in boiling benzine to turn it into the "yellow solution", bleaching clay is added to the solution, and then it is filtered to turn it into the "white solution".

Finally, the white solution is steam-distilled to evaporate the solvent. The purified gutta precipitates on the surface of water as shown in Figure 59. The wet white gutta which is soft but not sticky is taken out and passed over the hot roller to obtain dry "white gutta", as shown in Figures 60 and 61.

Figure 59 Purified gutta precipitating in the vapour-distillation vessel.

Figure 60 Taking out of white gutta.

Figure 61 Hot drying roller and dried white gutta (purity>98.5%).

The white gutta is weighed, moulded, canned and labelled as shown in Figures 62, 63 and 64.

Figure 62 Weighing white gutta (still warm and soft) and placing in the mould.

Figure 63 Moulds and press machine (left) and canning apparatus. (right)

Figure 64 Moulded white gutta (left) and labelling (right).

Benzine recycling system

The factory is equipped with a benzine recycling plant shown in Figures 65 and 66.

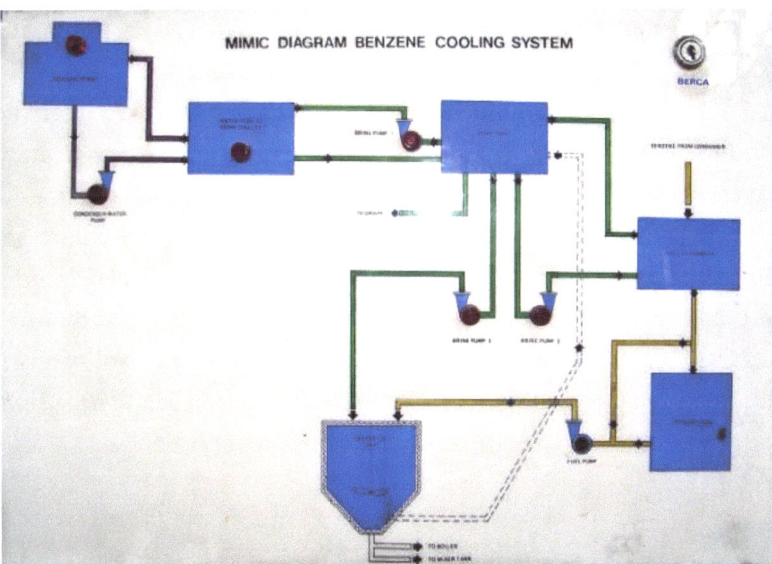

Figure 65 Panel to show the benzine recycling system.

Figure 66 The benzine recycling plant.

When stayed at Bogor, the present writer questioned whether the abovementioned process needed to be reviewed in the light of the twentieth century knowledge of macromolecular science. The late Dr. Sutrisno Budiman told him that in the past he had conducted various experiments on the extraction of gutta from the leaves, the use of modern absorbents to whiten the gutta, the recovery of gutta from the white solution, etc. and concluded that the methods and the total system contrived and designed hundred years ago were perfect, nothing to be modified. In fact, the use of benzine, which is much cheaper than benzene, alcohol, acetone, etc. and easily available, for both extracting resin and dissolving gutta at low and high temperatures, respectively, was most reasonable.

Chapter 8
Gutta Percha Conference 2007

A Gutta Percha Conference was held on February 15-16 2007, organised by PT Perkebunan Nusantara VIII and the present writer had an honour to participate in it. The conference consisted of a meeting at the corporation's lodge of Gunung Mas, Cisarua, Bogor (First day) and a visit to the Gutta Percha Plantation and Factory, Cipetir, Sukabumi (Second day).

The lodge located in the midst of Gunung Mas tea plantation which covered a northern hillside of Mt. Gede/Pangrango was a cosy place and the session was held in a large hall, shown in Figure 67 and Figure 68. A true *rijsttafel*[118] was served after the session.

Figure 67 A view from the Lodge of Gunung Mas Tea Plantation in which Mt. Salak is seen in the distance across the Bogor Valley.

Figure 68 A scene of the meeting, chaired by Dr. Suharto Honggokusumo (front table, dark-haired, dark shirt).

In the introductory presentation the present status of the Sukabumi Plantation was given. With regard to "The potentiality of the plantation in Sukabumi Regency", the slide (Figure 69) showed that almost one quarter of the total area of the regency (101,213/412,799 ha/ha) is occupied by plantations, in which the governmental plantation shares 22,695 ha. About "The gutta percha cultivated plantation (Figure 70)", it was explained that the area was reduced to 417 ha in recent years but the annual production of white gutta in 2003 – 2006 amounted, 5,076, 919, 1,582, and 2,032 kg, conducted on make-to-order basis.

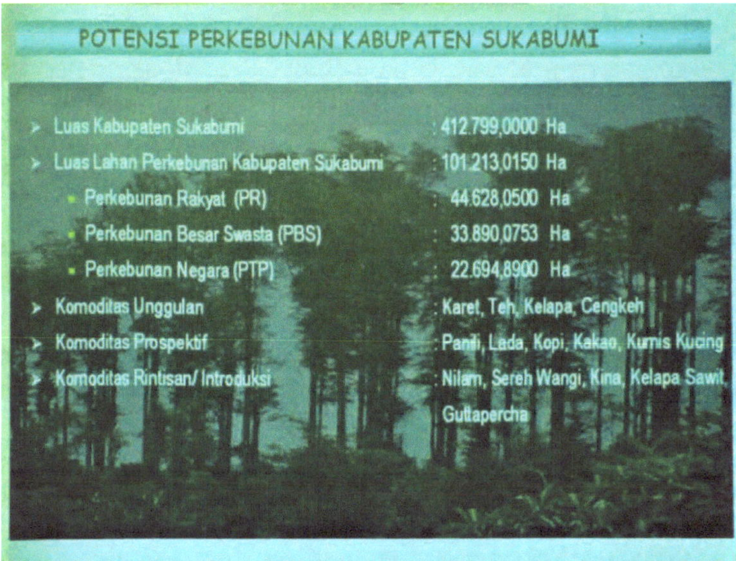

Figure 69 A slide to show the potentiality of plantation in Sukabumi Regency.

Figure 70 A slide to show the production of gutta percha in recent years.

To conclude the meeting, the Governor of Sukabumi Regency promised to support the gutta percha plantation/factory to be maintained in operational conditions and consider preserving the area as a historical asset or a technological heritage, creating a campsite in the plantation area and providing a cable car from the foot of the mountain.

According to Dr. Uhendi Haris, the former director of the Bogor Research Station for Rubber Technology, the Cipetir Plantation/Factory still operates today to meet the demands of gutta percha, although the idea to preserve the area as a historical asset has not been realised.

Figure 71 is just for the sake of nostalgia of the present writer to remember the late Dr. Augustine Francis Sutrisno Budiman, the promoter of the conference at Gunung Mas and Cipetir.

Figure 71 The late Dr. Augustine Francis Sutrisno Budiman on the occasion of conference at Cipetir.

Chapter 9
Recent News

According to BBC News 10 May 2013,[119] "About 40 large blocks of a rubber-like substance (Figure 72), believed to be from a shipwreck in the Atlantic Ocean, have washed up on European coasts. The pieces of gutta percha have been found on beaches in Cornwall, Devon, northern France and the Netherlands in the last year. ... As to the letters inscribed on the block, it said, 'TJIPETIR' is believed to be the name of a 19th Century rubber plantation".

Figure 72 A specimens housed in the Porthcurno Telegraph Museum, Cornwall.[120]

BBC News followed the details on 1 December 2014.[121] In the summer of 2012, a lady who was walking her dog along a beach near her home in Newquay, Cornwall, when she spotted a black tablet on the sand, made of something resembling rubber. By December 2013, more than eighty pieces of the same tablet were found washed ashore around the coast of the Great Britain and northwest European Continent (Figures 73 and 74). On the tablets were inscribed 'TJIPETIR', indicating they were gutta percha blocks produced at Tjipetir (Cipetir), West Java.

Figure 73 Mrs. Tracey Williams who first discovered blocks in the summer of 2012 and Mr. Jose de Cora who found a block in San Cibrao, northwest Spain in August 2013. BBC News.[122]

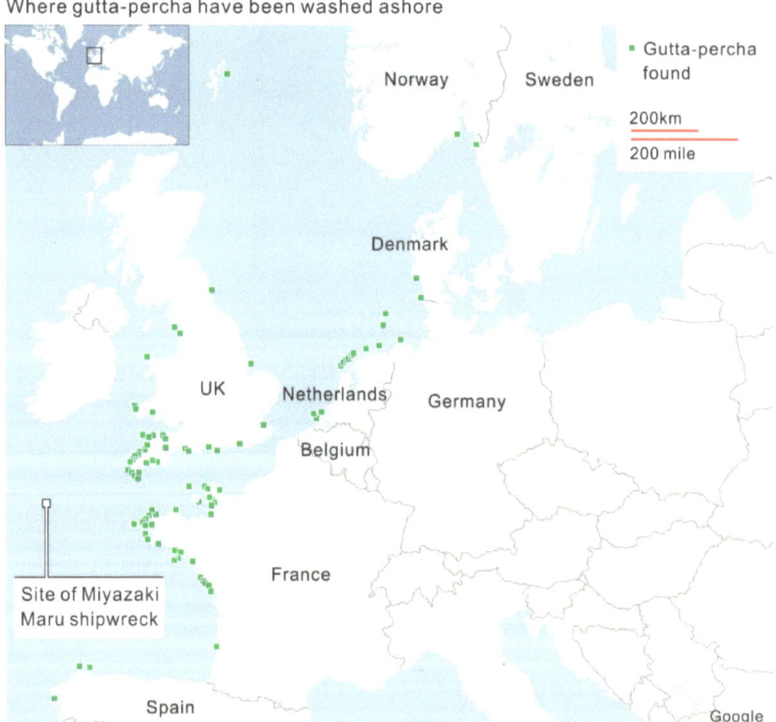

Figure 74 The sites of Tjipetir gutta percha discovered. BBC News.[123] Redrawn on CorelDRAW as similar as possible to clarify the details.

As the result of research, it turned out that those blocks had flown out of SS Miyazaki-Maru (Figure 75), a NYK's ocean liner of 8,520 tons, built in 1909 in Kobe, sunk on the 31 May 1917 during her voyage from Yokohama to London (via Singapore) by German submarine U-88 together with her eight passengers in the sea, 150 miles (241.5km) west of the Scilly Isles.

Figure 75 SS Miyazaki-Maru.[124]

Figures 76 and 77 shows old photographs of Cipetir Plantation obtained from the collection of Tropenmuseum.

Figure 76 A view of the Tjipetir factory (1920-1930),[125] and the stack of gutta-percha plates labelled "TJIPETIR" (ca. 1894-1929).[126]

Figure 77 The Tjipetir factory of old days. (1) Harvested Palaquium oblongifolium leaves (ca. 1920-1930),[127] (2) Cooking and separating process (ca. 1920-1930),[128] (3) Press process (Year unknown),[129] (4) The interior of new laboratory (ca. 1920-1930).[130]

The interior of the factory looks similar as it is today, although in the past steam power was transmitted from the line shaft to all machines, whereas presently every machine is provided with an electric motor. The press machine with a rectangular mould was different from that of today with a circular mould. The product was shipped probably not sealed in a can unlike today.

Appendix 1:
A Brief History of Xylonite (Celluloid)

(1) Introduction

The key to the development of xylonite (called Celluloid in America) rests on the observation of a phenomenon that when ca. 1860 Alexander Parkes mixed nitrocellulose (also called pyroxylin) with camphor, the fibrous substance turned into a thermoplastic which could be made into any shape at elevated temperatures around 75°C. At the same time, nitrocellulose, which was of use as gun cotton, became less combustible! The present writer would conjecture that the finding resulted not from logical thought but from try and errors.

Let us briefly look back the history of the nitration of cellulose in chronological manner with reference to literature.[131] Earlier in 1832, Henri Braconnot found that starch and similar bodies were rendered highly combustible through treatment with nitric acid. In 1838, Théophile-Jules Pelouze took the subject up and treated cotton, paper, and other vegetable substances with nitric acid. In 1843, Jean-Baptiste nitrated paper and obtained nitramidine, as he called, in the form of cartridges. The nitrated bodies produced in these early works lacked stability and uniformity. It was in 1846 that Christian Friedrich Schönbein and Rudolf Christian Böttger independently discovered a method to apply the mixed acid or the mixture of concentrated sulphuric and nitric acids, and successfully obtained homogenous nitrocellulose (Note that nitroglycerin was prepared in 1847 by Ascanio Sobrero by applying the same method).

In 1846, Louis-Nicolas Ménard and Florès Domonte discovered that cellulose nitrate could be dissolved in ether and called the solution diluted with ethanol "collodion". The collodion was applied next year by John Parker Maynard for surgery. In the 1850s,

S. Barnwell and A. Rollason proposed to use pyroxylin solutions as varnishes, paints, cements and coating materials for fabrics.

In 1851, Frederick Scott Archer and Peter Wickens Fry invented the collodion emulsion (wet plate) for photography. The quality of pictures obtained was better than that of calotype (silver salt applied to a paper, invented in the 1830s by William Henry Fox Talbot) and equivalent to that of Daguerreotype.[132] Figure A1-1 shows a contemporary scene of the collodion process.

Figure A1-1 The collodion process.[133]

The collodion process was useful until the gelatine dry plate was invented by Richard Leach Maddox in 1871.

(2) The development of Parkesine, Xylonite, Celluloid

Alexander Parkes (1813-1890), who invented the thermoplastic substance, named it as "Parkesine" and presented some samples and various products at The International Exhibition of 1862, London.[134] Figure A1-2 shows the page in the official catalogue, in which his products were registered just as "Patented Parkesine".[135]

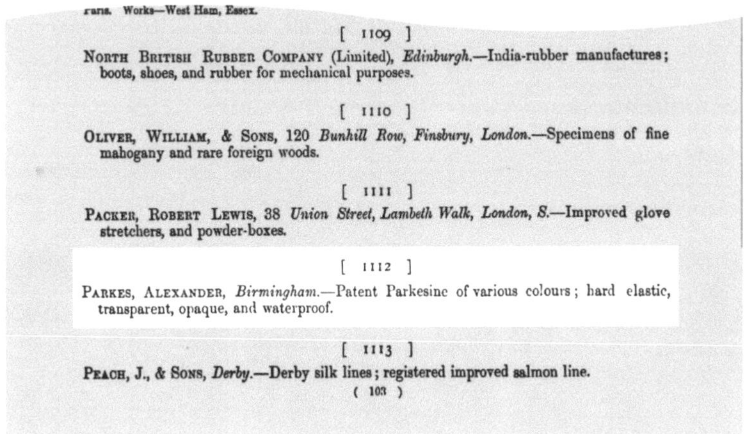

Figure A1-2 Alexander Parkes's "Parkesine" in the catalogue of The International Exhibition of 1862. Class IV: Animal and vegetable substances used in manufacturers; c. Vegetable substances used in manufacturers, &c.; Item number: 1112.[136]

In the Cassell's Illustrated Exhibitor for the same Exhibition[137], the composition of Parkesine was hidden, described as "a mixture of chloroform and castor oil".

> Among the most extraordinary substances shown is a new material called "Parksine," from the name of its discoverer, the product of a mixture of chloroform and castor oil, which produces a substance as hard as horn but as flexible as leather, capable of being cast or stamped, painted, dyed, or carved, and which, above all, can be produced, in any quantity, at a lower price than gutta percha.

The details were not disclosed either when Parkesine was exhibited in the Universal Exposition Paris1868.[138]

> La parkesine est le caoutchouc durci, mais par un autre procédé qui consiste à le mélanger avec de l'huile d'olive, du coton et autres matières analogues. ... (Parkesine is a hard rubber but made by another process consisting of the mixture of olive oil, cotton and the like. ...)

Figure A1-3 shows the Copper Medal, awarded to Parkes in The International Exhibition of 1862, and a parkesine-made plaque to commemorate the honour. Figure A1-4 shows various parkesine products.

Figure A1-3 Copper Medal, awarded to Alexander Parkes in The International Exhibition of 1862 (left), and a parkesine-made plaque to commemorate the award (right)[139]

Figure A1-4 Parkesine products ca.1860-1866, Curtain ring. Hair brush or mirror, Rectangular hair plate, Razor with parkesine handle, Fishing reel, .and Toothed gear wheel.[140]

The Parkesine Company was founded in 1866 but did not run well and within two years became liquidated. Figure A1-5 shows plaques to commemorate Alexander Parkes and the invention of Parkesine.

Figure A1-5 Plaques to commemorate Alexander Parkes[141] and the invention of Parkesine[142] at the old Birmingham Science Museum and the town hall of Hackney, London, respectively.

Then, two persons entered, Daniel Spill and John Wesley Hyatt, whose portraits are shown in Figure A1-6, together with Parkes's portrait.

Figure A1-6 Portraits of the three inventors, Alexander Parkes (1813-1890)[143], Daniel Spill (1832-1887)[144], John Wesley Hyatt (1837-1920)[145]

Daniel Spill, who worked with Parkes, took over the Parkes's patents and developed Ivoride or Xylonite, an improved form of Parkesine, which was plastic at about 75° C., by gelatinising pyroxylin in a solution of camphor dissolved in commercial grain alcohol. In 1877, he established The British Xylonite Company, sketched in Figure A1-7.

Figure A1-7 Daniel Spill's Ivoride Works, Homerton High Street, Hackney, c. 1880. Reproduced by London Borough of Hackney Archives (left),[146] and a company's products advertisement, May 1894 (right).[147]

Figure A1-8 shows various Xylonite products.

Figure A1-8 Various Xylonite products. (From left to right), Objects made of Ivoride and Xylonite, ca. 1869,[148] Xylonite hat brush - Sheffield, 1910, Hardy Brothers,[149] Xylonite Curves Set for Mechanical Engineers, Keuffel & Esser Co. New York,[150] Xylonite chess-set by British Chess Company Staunton Chessmen,[151] HMS Silver & Xylonite Boxed Set of Fish Cutlery.[152]

BX Plastics celebrated the company's centenary in 1977. Figure A1-9 shows the medallion for commemorating the Centenary of Xylonite with the Elephant and Tortoise trademark registered in 1893.

Figure A1-9 The Centenary Medallion of British Xylonite Company. This medallion is No 13 of a limited edition of 160 struck by the Birmingham Mint in 1977 and issued by British Industrial Plastics. By courtesy of Dr. John Goldsbrough, a friend of the present writer and former Technical Manager of the Chemicals and Cellulosics Division. Shown on the right is the Elephant and Tortoise logo engraved on a souvenir knife with xylonite handle which the present writer received during his visit to the Brantham Factory in Suffolk in 1980.

In the 1860s in America, John Wesley Hyatt acquired Parkes's patent and began experiments with cellulose nitrate, with the intention of making billiard balls, to substitute expensive ivory, and other articles. In 1872 Hyatt and his brother patented an improved manufacturing process and in the same year founded the Celluloid Manufacturing Company. Thus "Celluloid" became the American name of the pyroxylin-camphor plastic. A lawsuit occurred between Spill and Hyatt but in 1870. The petition of Spill was finally rejected in November 1884,[153] so that neither side could have exclusive privileges.

Figure A1-10 shows an old sketch of the industrial production complex of The Celluloid Company, Newark, New Jersey[154] and a drawing for the Hyatts' patent on the "Improved pyroxylin manufacturing process", United States Patent No. 133,229.[155]

Figure A1-10 A sketch of the industrial production complex of the Celluloid Company (ca. 1890), Newark, New Jersey (left)[156] and a drawing for the Hyatts' patent on the "Improved pyroxylin manufacturing process", United States Patent No. 133,229.[157]

Figure A1-11 shows a billiard ball and various Hyatt Celluloid products.

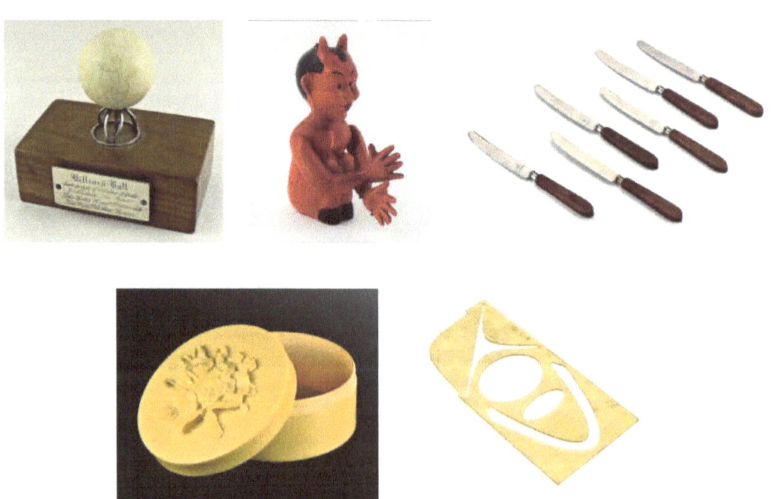

Figure A1-11 Various Hyatt Celluloid products. A billiard ball made in 1868,[158] A blow-moulded toy devil, Knives with celluloid-made imitation wood grain handle, A circular box, a confocal conic stencil (1900-1920).[159]

Xylonite (Celluloid) spread worldwide and became manufactured also in some other countries. In Japan, preparation of the pyroxylin-camphor mixture was first attempted in 1889. In the next decades a number of enterprises were established and, in 1919, eight major companies were united to form The Dai-Nippon Celluloid Company. The development of technology to prepare thin sheets and films expanded the application area.

Figure A1-12 shows a celluloid-surface slide rule that was indispensable for scientists and engineers of the present writer's generation before electronic calculators and small computers emerged in the late 1960s (The present writer remembers it took him three days and nights to obtain an electron density map and visualise the planer zig-zag structure of polythene molecule, for exercise, by conducting the calculation of the Fourier transform, using a slide rule and an abacus, on the basis of C. W. Bunn's structure factor data; later, the same calculation was done on a desk-top PC in mere minutes).

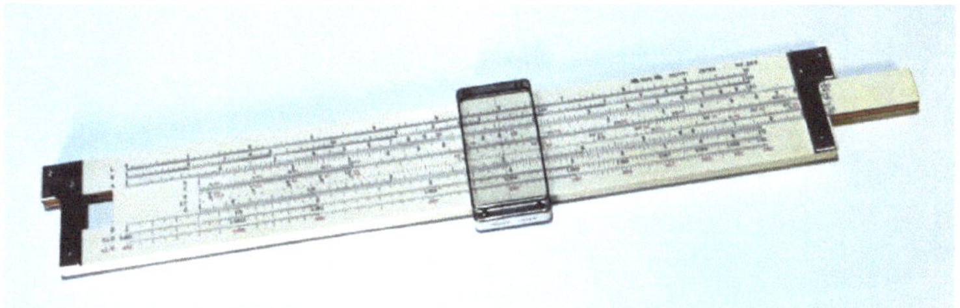

Figure A1-12 A double-sided, celluloid-surface Hemmi Bamboo Slide Rule No.259, the same type as the present writers had used.

(3) Application to Photographic and Motion Picture films

The application of transparent film for photography has a long and intricate story[160] but it was in 1888 that Eastman Kodak launched the original Kodak Camera loaded with a transparent roll film and boxed film for use in the camera, shown in Figure A1-13.

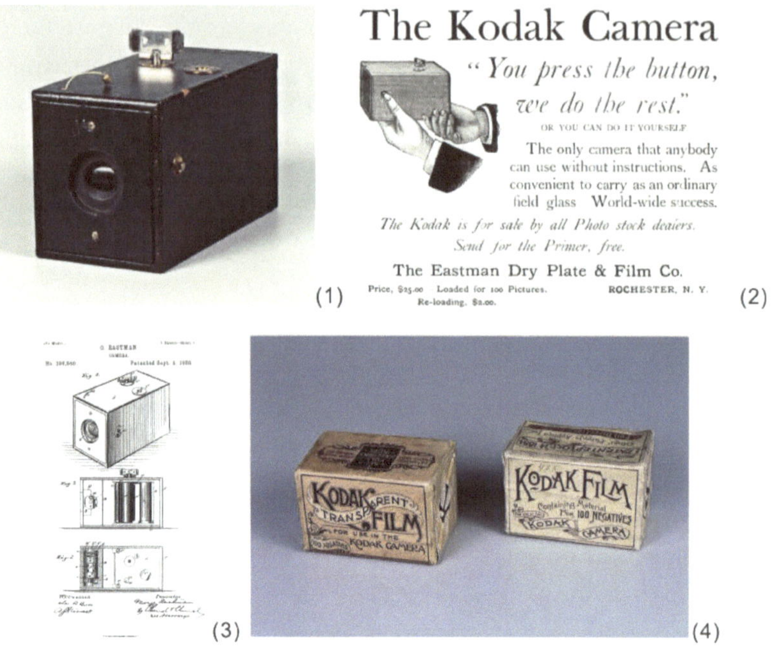

(1)   (2)

(3)   (4)

Figure A1-13 The Kodak Camera loaded with transparent roll film for 100 pictures. (1) The Camera,[161] (2) Advertisement of the camera,[162] (3) Drawings to show the details in U. S. Patent No. 388,650,[163] (4) Boxed roll films dated 1888 and 1889.[164]

Later in 1912, Eastman Kodak marketed sheet films which were useful for recording X-ray patterns. In 1925, Ernst Leitz launched the first 35 mm film camera, called "Leica 1".

Around that time, efforts to reproduce the image of moving bodies were made by a number of engineers. Figure A1-14 is for reference to show two prototype motion picture apparatuses, the zoopraxiscope disc and the electrotachyscope, invented by Eadweard Muybridge and Erwin F. Faber (ca. 1893) and Ottomar Anschütz (1886-1894), respectively.

Figure A1-14 Prototype motion picture apparatuses.
Left: A zoopraxiscope disc by Eadweard Muybridge and Erwin
F. Faber (ca. 1893)[165] Right: Electrotachyscope, invented by
and Ottomar Anschütz (1886-1894).[166]

In 1895 France, the Lumière brothers (Auguste and Louis) invented Cinematograph, for which a strip of nitrocellulose-based film was used, as shown in Figure A1-15. Since the film was said to be inflammable, the composition of the film was probably similar to that of xylonite or celluloid.

Figure A1-15 Lumiere Cinematograph, 1895.[167;168]

Soon after then, rolled films for motion pictures were developed as examples are shown in Figure A1-16.

Figure A1-16 Celluloid roll films. Left: 35mm black and white roll film invented by William Kennedy Dickson and manufactured by Edison company/Eastman Kodak.[169] Right: Hannibal Goodwin's roll film 1887 (used in Thomas Edison's Kinetoscope).[170]

Thus, the market of celluloid greatly expanded in accordance with the development of photography and cinematography industries.

(4) Decline

Because the flammability of nitrate-based films was inevitable, Eastman Kodak developed cellulose acetate film in the 1920s, as a substitute, and eventually discontinued the production of cellulose nitrate film in 1951. Other companies followed suit. Other xylonite- and celluloid-made items, save table tennis balls, some glass frames and stationaries, had gradually disappeared by the 1970s, replaced with synthetic plastic products. In such circumstances, in 2005, BX Plastics terminated the production of Xylonite, at the company's factory at Brantham, Suffolk. In 2014, the International Table Tennis Federation (ITTF) made a decision to mandate the use of plastic balls in all international events, instead of the traditional celluloid balls.

Consequently, celluloid is now only of use for some glass frames, the stem of vintage fountain pens and other minor items.

(5) The reason why pyroxylin mixes homogeneously with camphor

A polymer chemist, the present writer has long wondered why pyroxylin is miscible with camphor without regard to their composition. Figure A1-17 shows the result of infrared absorption measurement on a dried camphor-collodion mixture conducted courtesy of Prof. K. Tashiro, Toyota Technological University in 2018. It was observed that the absorption band of the >C=O group at 1743 cm^{-1} had shifted by 10 cm^{-1} towards the lower wavenumber side, revealing the association of >C=O groups of camphor with the –NO2 groups in pyroxylin. The specimen employed for reference was an old triangle ruler, which was probably made in England in the early 20th Century, found in the present writer's desk drawer. The measurement simultaneously proved that the ruler, shown in Figure A1-18, was made of xylonite.

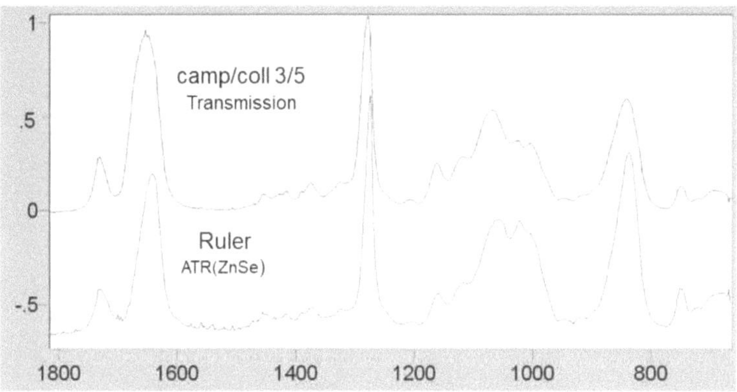

Figure A1-17 FTIR (Fourier-transfer infrared) spectra of a dried camphor/collodion 3/5 mixture and an old xylonite triangle ruler.

Figure A1-18 A triangle ruler found in the present writer's desk-drawer (proved to be made of xylonite from the above spectra).

(6) The Production of Camphor

Let us review the history of camphor, as it was the important component of Xylonite or Celluloid. Camphor trees (*Cinnamomum camphora*), which was native to in Asia, viz. in Annam, South China, Taiwan, and the southern part of Japanese archipelago, was used as a medicine and cosmetic ever since ancient time and traded to other parts of Asia and Europe. Figure A1-19 shows the botanical illustration, a tree and the wood chip of a camphor tree, and Figure A1-20, the distribution of trees in the world.

Figure A1-19 Cinnamomum camphora. (1) Botanical illustration,[171] (2) An old camphor tree in the Koishikawa Botanical Garden, Tokyo,[172] (3) Wood chips of camphor tree.[173]

Figure A1-20 The world distribution of *Cinnamomum* camphora.[174] Redrawn by the present writer on CorelDRAW.

Ever since the seventeen century, camphor was a major trade item of Japan (next to silver and gold), exported to China and Europe by the Dutch East India Company). Originally camphor was obtained by simply cooking the tips of camphor wood as shown in Figure A1-21.

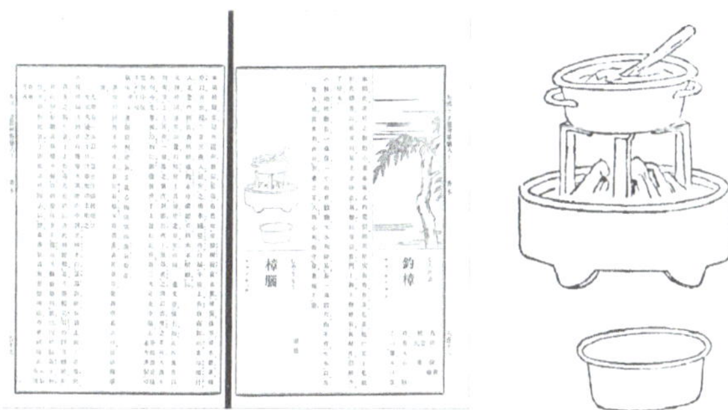

Figure A1-21 Description on camphor in a Japanese book, published in 1712, in which old Chinese literature was referred to (left) and the enlarged sketch of the collecting method (right).[175]

In the primitive steam-distillation method (below), only the solidified camphor was collected and the so-called camphor oil which contained significant amount of camphor was drained (See Figure A1-22)

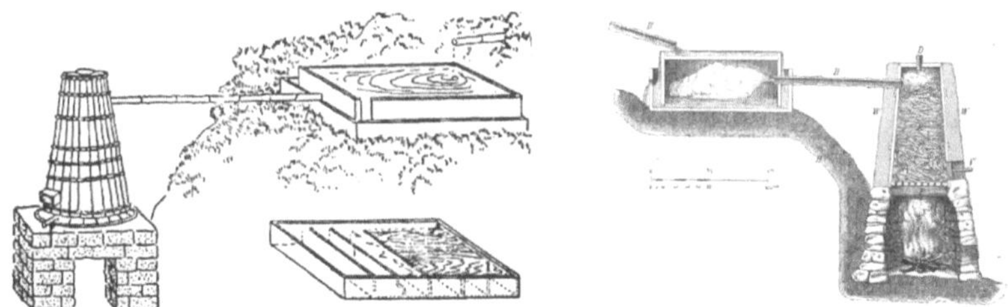

Figure A1-22 A sketch of camphor-still in Tosa, Japan, in 1880s (left)[176] and the details of the same device.[177]

Around the turn of the century to the 20th, steam-distillation process was improved to collect camphor as well as camphor oil, as shown in Figure A1-23. From 1,000 kg wood (containing 50% water), 8 kg camphor and 4 kg camphor oil were obtained.

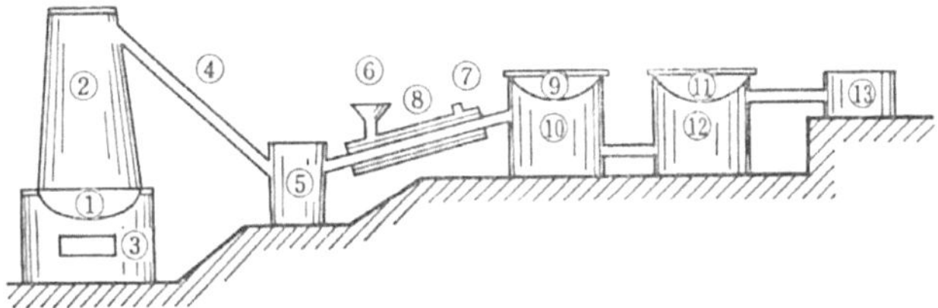

Figure A1-23 Camphor and camphor oil manufacturing apparatus, used in some places in Japan and Taiwan until ca.1960.[178]

By means of the fractional distillation of the camphor oil, shown in Figure A1-24, 40 % camphor was recovered and 26 % white oil and 24% black oil were obtained.[179]

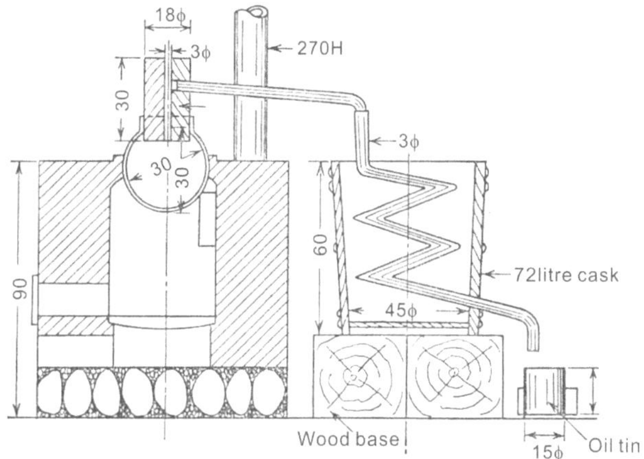

Figure A1-24 The first camphor oil still of modern design built in Japan in 1879 (Unit: cm, converted from the traditional Japanese unit).[180]

Modern fractionation towers became common in later years.

In 1903, the synthesis of camphor was accomplished by Gustav Komppa, a remarkable achievement in the history of organic chemistry, whose portrait is shown in Figure A1-25, although natural camphor was competitive against synthetic camphor.

Figure A1-25 Gustav Komppa (1867-1949)[181]

The demand for camphor drastically decreased in the 1970s with the decline of xylonite and celluloid. Nowadays the application of camphor is limited in the traditional cosmetic, medical and sanitary areas, besides a minor amount of xylonite (celluloid) which is still produced for specific purposes mentioned above.

Appendix 2:
Thermoplastic Aspects of Amber and Copal

Amber is a term for fossilised resin from extinct coniferous trees found in many places in the world. Although the initial substance is assumedly polylabdanoid from labdatriene precursors, the structure is considered to have become more heterogeneous, modified by the occurrence of crosslinking, isomerisation and cyclisation while buried under the soil for long period (>40 million years). Consequently the properties as well as the appearance are diverse depending on its origin. Needless to say, amber has been valued as jewellery or ornamental material since ancient times and made into decorative objects, ranging from small caskets to the spectacular Amber Room in the Catherine Palace, Saint Petersburg. Amber was also used for folk medicine.

Copal may be defined as a collective term for resins from various tropical trees modified in a shorter period (3-4 million years). The trees included *Protium copal* from Mexico (from which the word "copal" originated), *Guibourtia copallifera* from Congo, *Agathis australis* from New Zealand, etc. The properties of copal are also various.[182; 183]

Table A2-1 shows the summary of DSC measurement conducted on various amber and copal specimens.

Table A2-1 Glass transition temperature of fossilised resins determined by TOPEM.[184]

Material	Tg (1st heat) °C	Tg (2nd heat) °C	ΔH (postcure) J/g	Age 10⁶ Yrs.
Dominican amber	163	166	46	20?
Baltic amber	143	152	59	40
Kauri gum (old)	142	147	42	40
North Sea amber	110	120	19	-
Copal	82	101	57	1
Kauri gum (young)	74	96	51	0.01
Kauri gum (Philip.)	70	98	86	0.01
Colophonium	40	45	7	-
Tree resin	18	26	99	0

Amber is not truly thermoplastic but deformation and fusing is possible at high temperature under high pressure, although the fossilised and cross-linked resin does not turn fluid but decompose above certain temperatures (>300°C).[185] Figure A2-1 shows a bracelet of Amberide (pressed amber) and a cross pendant made of fused amber.

Figure A2-1 Heat-pressed amber products. Left: A bracelet made of pressed amber from Myanmar.[186] Right: a cross pendant made of fused amber.[187]

Copal has been widely applied for manufacturing varnish, lacquer, adhesives and fake amber products, besides traditional use as incense.

Taking advantage of low softening temperature or glass transition temperature, copal has long been used as a moulding material. In 1920s, ornaments of copal mixed with cellulosic tissue, pigments and other chemical ingredients, named Ebena, were developed at a Belgian company at Wijnegem,[188] as examples are shown in Figure A2-2.

Figure A2-2 Examples of Ebena products. Left: A tobacco jar, ca. 1925, Right: Octagonal bonbon box with gold leaf, ca. 1928.[189]

In New Zealand, the native Maori people used copal (Kauri Gum) for moulding such artefacts shown in Figure A2-3.

Figure A2-3 A pair of early 1900s kauri gum busts exquisitely moulded depicting a Maori chief with full face *moko* and a chiefteness with a *moko kauae*.[190]

Pressed copal is generally called Ambroid and widely used as a substitute for amber. Figures A2-4 shows a typical bead necklace sold on the internet. Figure A2-5 shows a meerschaum pipe with an ambroid stem which the present writer bought about twenty-five years ago in a famous tobacco shop in Amsterdam and added to his collection.

Figure A2-4 An ambroid bead necklace.[191]

Figure A2-5 A meerschaum pipe with an ambroid stem from the present writer's collection.

References and Notes

1 MacDonald, George. *The History of Gutta Percha Willie, the Working Genius*, Blackie & Son, London 1873.

2 Clarke, J. O. (Illustration by William Dalton), *Gutta-Percha, Its Discovery, History, Remarkable Properties, Vast Utility, and Application to Scientific and Ornamental Purposes, also Its Economy, and Importance as a Sanatory agent, with Instructions for Soling Boots and Shoes, Joining Driving Bands and Tubings, and for the Preventing Tooth-ache, by Stopping the Caries.* J. O. Clarke, London, 1849.

3 Green, Benjamin L. *Gutta percha, its discovery, history, and manifold uses.* Benjamin L. Green, London,1851.

4 A resin secreted by the female lac bug (Kerriidae Kerria lacca) on trees in the forests of India and Thailand. Consisting of hydroxy fatty acids, it is soluble in alkaline solutions and organic solvents, typically ethanol. Used for varnish, adhesives, etc.

5 Faraday, Michael. "XXV. On the Use of Gutta Percha in Electrical Insulation", *Philosophical Magazine and Journal of Science., 3rd Series* (1848 March): 165-167.

6 Robertson, J. C. (edit). *The Mechanics' Magazine, Museum, Register, Journal and gazette,* London 50 (1849 January 5 - June 30): 99-101.

7 Williams, C. Greville. *Chemical News* 2, no 45 (1860 October 13): 206.

8 Tilden, William A. "Note on the Spontaneous Conversion of Isoprene into Caoutchouc." *Proceedings of Birmingham Philosophical Society* 8 (1892): 182-183. The author wrote, "I was surprised a few weeks ago at finding the contents of the bottles containing isoprene from turpentine entirely changed in appearance. In place of a limpid colour less liquid, the bottle contained a dense syrup in which was floating several large masses of a solid of a yellowish colour. Upon examination, this turned out to be India-rubber." In the light of today's knowledge, the "India-rubber", presumably generated cationically by a small amount of some acidic contaminants, was probably not pure trans-1,4-poly-isoprene but a "copolymer" consisting of 1,4- and cyclized units. (Cf. e.g., Ouardad, Samira, Alain Deffieux, and Frédéric Peruch. "Polyisoprene synthesized via cationic polymerization: State of the art." *Pure and Applied Chemistry* 84, no. 10 (2012): 2065–2080).

9 Obach, Eugene F. A. "Cantor lectures: Gutta percha. November 29, 1897." *Journal of the Society of Arts* 46, no. 2354 (1897 December 31): 117-136.

10 Williams, Llewelyn. "Laticiferous Plants of Economic Importance V. Resources of Gutta-Percha-Palaquium Species (Sapotaceae)." *Economic Botany* 18, no.1 (1964, January - March): 5-26.

11 Burck,W., *Mededeelingen uit 's Lands Plantentuin. I: Rapport omtrent een onderzoek naar de getah-pertja produceerende boomsoorten in de Padangsche Bovenlanden,* Batavia Landsdrukkerij, 1884.

12 Burck, W. *Wandelingen door den Botanischen Tuin te Buitenzorg,* Batavia Landsdrukkerij, 1892.

13 See Ref. 11 above.

14 Isonandra gutta. http://www.plantillustrations.org/illustration.php?id_illustration=195618&SID

15 Palaquium oblongifolium, Burck. http://plantgenera.org/species.php?id_species=739183

16 Hevea brasiliensis. http://powo.science.kew.org/taxon/urn:lsid:ipni.org:names:349913-1

17 Tschirch, A.; Otto Oesterle. "Untersuchungen uber die Sekrete I. Studien uber die Guttapercha." *Archiv der Pharmazie. Zeitschrift des Deutschen Apotheker Vereins, Berlin* (1892): 641-670.

18 Gomez, J. B.; Narayanan R.; Chen, K. T. *Journal of Rubber Research Institute, Malaya* 23, no. 3 (1972): 193-203. (After Riches, J. P.; Gooding E. G. B. "Studies in the Physiology of Latex I. Latex Flow on Tapping - Theoretical Considerations." *The New Phytologist* 51, no. 1 (1952): 1-10.

19 See Ref. 9 above.

20 Sarawak: four Kayan natives collecting gutta percha from a tree. https://commons.wikimedia.org/wiki/File:Sarawak;_four_Kayan_natives_collecting_gutta_percha_from_a_t_Wellcome_V0037406.jpg.

21 File:Les merveilles de l'industrie, 1873 "Récolte de la gutta-percha dans une fôret de la malaisie". (4726556841).jpg

22 See Ref. 10 above.

23 See Ref. 9 above.

24 Ibid.

25 Oxley, Thomas. "Gutta percha," *Journal of Indian Archipelago and Eastern Asia. 1* (18471): 22-34.

26 Paul, B. H. (edit). *Industrial Chemistry: A manual for use in technical colleges or schools and for manufacturers etc., based upon a translation (partly by Dr. T. D. Barry) of Stohmann & Engler's German edition of Payen's 'Precis de Chimie Industrielle'.* Longmans, Green & Co., London, 1878.

27 Wray, jun, S. *Journal of the Society of Arts 33 (1884 November 21 1885 November 13).* This is the resume of article originally published in Journal of Straits Settlements Branch of the Royal Asiatic Society (1884).

28 See Ref. 26 above.

29 See Ref. 25 above.

30 See Ref. 26 above.

31 T. Oxley assumed (in note 25) that 69,180 trees must have been sacrificed to obtain 6,918 piculs 418,400 kg) of gutta percha in 1845-1847 (Av. 6.05 kg/tree).

32 See Ref. 9 above.

33 Tradescant, John. *Musaeum Tradescantianum: Or, a Collection of Rarities Preserved at South-Lambeth Neer London.* Grismond, 1656. https://books.google.co.jp/boo

34 Ibid.

35 Pinkerton, W. "Mazer Wood: Gutta Percha." Notes and Queries: A medium of inter-communication for literary men, artists, antiquaries, genealogists, etc. no.74 (1851 Maech 29). https://www.gutenberg.org/files/23282/23282-h/23282-h.htm

36 See Ref. 9 above.

37 Allen. "Literary & Scientific Intelligence." *Allen's Indian Mail, and register of intelligence of British and foreign India, China, all parts of the East,* 6 (1848 January-December): 504.

38 Sir Jose d'Almeida - A Catholic Pioneer in every sense. https://history.catholic.sg/sir-jose-dalmeida/

39 Montgomerie - Plastics Historical Society. http://plastiquarian.com/?page_id=14213

40 See Ref. 3 above.

41 Ibid.

42 Ibid.

43 Ibid.

44 Hard Rubber Buttons & Some Look-A-Likes. http://www.vintagebuttons.net/rubber.html

45 c1870 French Gutta Percha Plastic Case with Napoleon Portrait. https://paradeantiques.co.uk/antiques/other/c1870-french-gutta-percha-plastic-case-with-napoleon-portrait

46 A gutta percha hand mirror. https://www.andreburgos.com/products/a-gutta-percha-hand-mirror

47 Victorian Gutta Percha Chessmen. https://chessantiques.com/product/victorian-gutta-percha- chessmen/

48 *Official Descriptive and Illustrated Catalogue of the Great Exhibition of the Works of Industry of All Nations, 1851, Vol. 2,* Spicer brothers, London, 1851. https://catalog.hathitrust.org/Record/001511518.

49 Ibid.

50 *The International Exhibition of 1862: The illustrated catalogue of the Industrial Department, British Division. Vol I, Printed for Her Majesty's Commissioners, London 1862*. Class IV: Animal and vegetable substances used in manufacturers. Sub-Class A. Oils, Fats, and Wax, and their Product, pp.313-314.

51 Ibid..

52 Brannt, William T. *India Rubber, Gutta-Percha, and Balata: Occurrence, Geographical Distribution, and Cultivation of Rubber Plants; Manner of Obtaining and Preparing the Raw Material, Modes of Working and Utilizing Them, and Statistics of Commerce, Second Edition*, Scott, Greenwood & Sons, London 1910.

53 harles Bright, Submarine Telegraphs, Crosby Lockwood & Son, London, 1898. Cambridge University Press, 2014. 292-292.

54 See Ref. 52 above.

55 Signer, Charles, et al. (edit), *A History of Technology Vol.4, The late nineteenth century, c. 1850-1900*, Oxford University Press, 1958.

56 Williams, C. Greville. "XXIII. On lsoprene and Caoutchouc. *Journal of Chemical Society*." 15 (1862): 11-20. Abstracted from the *Philosophical Transasctio DS*. 1860: 241.

57 See Ref. 26 above.

58 Willmarth, Geo. E. "Purification of commercial gutta percha, and preparation of liquor gutta-percha." *Pharmaceutical Journal and Transaction, London, 6* (1875): 409-410.

59 How, W. Storer. "The evolution of artificial tooths and cap crowns and bridge-work." *The Dental Advertiser Vol. 17, no. 3*, Buffalo, N. Y. (1886): 87-95.

60 Jacob, Henry Long. "The toughness of gutta percha when treated with chloroform," *British Journal of Dental Science, London* 16, no. 199, (1873).

61 See Ref. 59 above.

62 Gutta percha bottle, 1974. Credit: Science Museum/Science & Society Picture Library.
https://webarchive.nationalarchives.gov.uk/20170423085616/http://www.ingenious.org.uk/See/Tradeandindustry/Textileindustry/?print=1&target=SeeMedium&ObjectID={C7108460-141B-C502-5366-65CA1DD98FCE}&s=S1&source=Search&SearchCategoryID={C30DE785-2657-4A36-A5A6-000000036641}&viewby=images&

63 The Times. "Golf chiefs take swing at big hitters to make game more fun." Saturday November 25 2017, 12.01am GMT, The Times https://www.thetimes.co.uk/article/golf-chiefs-take-swing-at-big-hitters-to-make-game-more-fun-2lmrw7qn9

64 Ibid.

65 A friend of progress. "German experiments on the insulating properties of gutta percha-covered electric wires by Mr. Werner Siemens." *The Mechanics' Magazine, Museum, Register, Journal and gazette, January 5th June 30th, 1849, Edited by J. C. Robertson, Vol. L*, London, 1850. p.99-101.

66 See Ref. 5 above.

67 Reif-Acherman, Simon. "Ernst Werner Von Siemens and the Early Evolution and Diffusion of Electric Telegraphy." *Proceedings of the IEEE 105, no. 11 (2017 November)*: 2274 - 2284. DOI: 10.1109/JPROC.2017.2752583

68 Murray, John. *German experiments on the insulating properties of gutta percha covered electric wires by Mr. Werner Siemens, Scientific & Technical Papers of Werner von Siemens* (Translated from the Second German Ed.), Vol. II: Technical papers, London , 1895, pp.27-30.

69 See Ref. 55 above.

70 Ibid.

71 Torchiana, H. A. van Coenen. *Tropical Holland, An Essay on the Birth, Growth and Development of Popular Government in an Oriental Possession,* University of Chicago Press, 1921.

72 Sample of submarine telegraph cable, 1866. https://collection.sciencemuseumgroup.org.uk/objects/co33695/sample-of-submarine-telegraph-cable-1866-cable

73 See Ref. 53 above.

74 C. William Siemens, in: *The record of the International Exhibition 1862*, W. Mackenzie, London 1862, p.529.

75 The Reels of Gutta-percha Covered Conducting Wire Conveyed into Tanks at the Works of the Telegraph Construction and Maintenance Company, at Greenwich 1865–66 by Robert Charles Dudley. https://www.metmuseum.org/art/collection/search/383811?searchField

76 See Ref. 9 above.

77 India Rubber, Gutta Percha and Telegraph Works Company. https://en.wikipedia.org/wiki/India_Rubber_Gutta_Percha_and_Telegraph_Works_Company

78 Ibid.

79 See Ref. 7 above.

[80] See Ref. 56 above.

[81] Bunn, C. W. "Molecular structure and rubberlike elasticity. II. The stereochemistry of chain polymers." *Proceedings of the Royal Society, London A.* 180 (1942):40-66.

[82] Ibid.

[83] Storks, K. H. "An Electron Diffraction Examination of Some Linear High Polymers." *Journal of American Chemical Society* 60 (1938,): 1753.

[84] Keller, A. "A note on single crystals in polymers: Evidence for a folded chain configuration." *Philosophical Magazine* 2 (1957): 1171-1175.

[85] Till Jr., P. H. "The growth of single crystals of linear polyethylene." *Journal of Polymer Science A* 24 (1957): 301-306

[86] Hochpolymeren." *Zeitschrift für Naturforschung* 12a (1957): 753-754.

[87] See Ref. 83 above.

[88] Kohji, Tashiro; Kobayashi, Masamichi; Tadokoro, Hiroyuki. "Elastic Moduli and Molecular Structures of Several Crystalline Polymers, Including Aromatic Polyamides." *Macromolecules* 10, no.2 (1977): 413-420. doi.org/10.1021/ma60056a033

[89] Ratri, Paramita Jaya; Tashiro, Kohji; Iguchi, Masatoshi; "Experimentally- and theoretically-evaluated ultimate 3-dimensional elastic constants of trans-1,4-polyisoprene α and β crystalline forms on the basis of the newly-refined crystal structure information." *Polymer* 53 (2012): 3548-3558.

[90] Ratri, Paramita Jaya; Kohji Tashiro. "Phase-transition behavior of a crystalline polymer near the melting point: Case studies of the ferroelectric phase transition of poly(vinylidene fluoride) and the β-to- α transition of trans-1,4-polyisoprene." *Polymer Journal* 45(2013): 1107-1114.

[91] Ibid.

[92] Szczepanowska, Hanna M.; Horie, C. Velson. "Gutta Percha, Rubber or Balata? Establishing Analytical Markers for Identification of Cultural Heritage Materials", *Microscopy and Microanalysis* 24 (Suppl 1) (2018): 2162-2163.

[93] Ibid.

[94] See Ref. 9 above.

[95] See Ref. 10 above.

[96] See Ref. 10 above.

[97] See Ref. 9 above.

[98] See Ref. 11 above.

[99] See Ref. 12 above.

[100] See Ref. 9 above.

[101] See Ref. 10 above.

[102] Burns, Russell M.; Honkala, Barbara H. *Silvics of North America Agriculture Handbooks : Volume 2. Hardwoods*, US Department of Agriculture, 1990, p.455-459.

[103] Brydson, J. A. *Plastics materials Fifth edition*, Butterworth, 1989. p.813.

[104] Manilkara bidentata (A. DC.) A. Chev. http://www.plantillustrations.org/illustration.php?id_illustration=706 32&id_taxon

[105] Manilkara Zapota. http://www.plantillustrations.org/species.php?id_species=641326& mobile

[106] Laan, Joseph W. Vander. *Production of gutta-percha, balata, chicle and allied gums*. Government Printing Office, Washington, 1927. p.26-48.

[107] *British Chemical Abstracts: A. Pure chemistry, Bureau of Chemical Abstracts*.1930, p.610. https://books.google.co.jp/books?id=qsY_AQAAIAAJ&q

[108] See no.92 above.

[109] Forrest, M. J., *Rubber Analysis - Polymers, Compounds and Products Smithers*. Rapra Press, 1996. p.48.

[110] Andrews, S., "Tree of the Year : Eucommia ulmoides." *International Dendrology Society Yearbook*. 2008. p.16-37.

[111] Ibid.

[112] Ibid.

[113] Nakazawa Y.; H. Uji. "Development of Tu-chung elastomers with inedible biomass." http://www.osaka-u.info/pdf/2015/ 07/top_article_2.pdf [Japanese lang]

[114] Hitachi Zosen Corp. "What is Tu-chung elastomer?" https://www.hitachizosen.co.jp/products/products003.html [Japanese lang]

[115] Ibid.

[116] Yoshikawa, R.; Shida K.; and Harima, H. "On the synthetic trans-1,4-poly-isoprene." *Journal of the Society of Rubber Industry, Japan*. 51 (1978): 20-24. [Japanese lang].

[117] Yoshikawa, R. "Trans-1,4-poly-isoprene." *Journal of the Society of Rubber Industry, Japan*. 57 (1984): 723-727. [Japanese lang]

[118] Rijsttafel, a Dutch-style Javanese meal, served in buffet-style, which is said to have originated in old days in plantations that were located remote from towns.

119 BBC. "Shipwrecked gutta percha blocks wash up in South West, 10 May 2013." https://www.bbc.com/news/uk-england-22478533

120 Ibid.

121 BBC. "Tjipetir mystery: Why are rubber- like blocks washing up on beaches?" by Mario Cacciottolo, BBC News, 1 December 2014. https://www.bbc.com/news/magazine-30043875

122 Ibid.

123 Ibid.

124 SS Miyazaki-Maru. https://www.wrecksite.eu/wreck.aspx?138262

125 Guttaperchafabriek van de plantage Tjipetir 1920-1930, TM-60016732. https://collectie.wereldculturen.nl/#/query/18b2fd48-078e-4973-995a-7a93f2481570

126 Aflevering produkt in blokken van onderneming 'Tjipetir' 1894-1929. TM-10012863. https://collectie.wereldculturen.nl/#/query/afdb1f76-7e06-4097- 81a3-e9710d0e5f31

127 Tjipetir, de bladzolder van de nieuwe guttapercha fabriek ca. 1920-1930. TM-60020168. https://collectie.wereldculturen.nl/#/query/c8bb80ef-cca9-4f7f- a80a-c71428611e7f

128 Tjipetir, fabriekshal met kookpannen en afscheidingsbakken voor de verwerking van guttapercha ca. 1920-1930. TM-60020166. https://collectie. wereldculturen.nl/#/query/29aef745-8d3c-4420-8ca2-0748ceb555d9

129 Tjipetir, arbeiders bedienen de machines die de guttapercha wassen en persen ca. 1920-1930. TM-60020167. https://collectie.wereldculturen.nl/#/query/ 82aba8e2-b8c0-4ab2-95dd-5c78a04ae5e3

130 Interieur van het laboratorium van de nieuwe guttapercha-fabriek bij Tjipetir, West-Java ca. 1920-1930. TM-60020169. https://collectie.wereldculturen.nl/#/query/76c8ad35-df2b-44f9-9be8-4ea9c04ba723

131 Schltpciluts, Robert C. "The evolution of smokeless powder", *J. Soc Chem Ind, June 29*, 1895, pp556-559, etc.

132 http://www.collodion.org/q&a.html

133 Ibid.

134 Krebs, Robert E. (ed.), Rae Dejur (illust). *Encyclopedia of Scientific Principles, Laws, and Theories. Volume 2: L–Z*, Greenwood Press, London 2008, p.427-428, etc.

135 *The International Exhibition of 1862: The illustrated catalogue of the Industrial Department, British Division – Vol I,* Printed for Her Majesty's Commissioners, London 1862. Class IV: Animal and vegetable substances used in manufacturers. c. Vegetable substances used in manufacturers, &c.

136 Ibid.

137 *Cassell's Illustrated Exhibitor; containing about three hundred illustrations, with letter-press descriptions of all the principal objects in the International Exhibition of 1862,* p.43.

138 Chevalier, M. Michel, *Exposition Universelle De 1867 A Paris, Ripports du Jury international, Tome Sixième, Groupe V - Classes 41 a 43,* Paris, de Paul Dupont , 1868

139 https://collection.sciencemuseum.org.uk/search?q=Parkesine.

140 https://collection.sciencemuseumgroup.org.uk/search?q=Parkesine

141 https://commons.wikimedia.org/wiki/File:Alexander_Parkes_Blue_Plaque.jpg

142 https://www.plaquesoflondon.co.uk/locations/alexander-parkes-e9/

143 http://www.robinsonlibrary.com/technology/technology/biography/parkes.htm.

144 http://www.robinsonlibrary.com/technology/chemical/biography/spill.htm.

145 http://www.wikiwand.com/en/John_Wesley_Hyatt.

146 https://rawmaterials.bowarts.org/collection/british-xylonite- company-founded-1877/

147 https://www.gracesguide.co.uk/British_Xylonite_Co

148 Susan Mossman, "Early plastics: perspectives 1850–1950", Ferrum 89, 2017, p.14-24.

149 http://www.robinsonlibrary.com/technology/chemical/biography/spill.htm.

150 https://www.shopgoodwill.com/Item/51442150

151 https://chessantiques.com/product/british-chess-company-staunton-chessmen/.

152 The set consists of twelve knives and forks in velvet lined box.. http://www.lawsons.com.au/asp/fullCatalogue.asp?salelot=7927A+++++10+&refno

153 Blatchford, Samuel. *Reports of Cases Argued and Determined in the Circuit Court of the United States, For the second Circuit,* Volume XXII. Baker, Voorhis & Co., Publishers, 66 Nassau Street, New York 1885.

154 https://www.wikiwand.com/en/Celluloid

155 https://pdfpiw.uspto.gov/.piw?Docid=00133229&homeurl

156 https://www.wikiwand.com/en/Celluloid

157 https://pdfpiw.uspto.gov/.piw?Docid=00133229&homeurl

158 http://americanhistory.si.edu/collections/search/object/nmah_2947.

159 https://collection.sciencemuseumgroup.org.uk/

160 Raymond Fielding, A technological history of motion pictures and television, Univ. California Press 1984.

161 https://notquiteinfocus.com/2014/04/23/a-brief-history-of-photography-part-6-kodak-the-birth-of-film/

162 Ibid.

163 https://commons.wikimedia.org/wiki/File:George_Eastman_patent_no_388,850.png

164 https://www.eastman.org/two-rare-rolls-early-kodak-film-acquired-george-eastman-museum

165 https://en.wikipedia.org/wiki/Zoopraxiscope

166 https://en.wikipedia.org/wiki/Electrotachyscope

167 https://history.com/news/the-lumiere-brothers-pioneers-of-cinema

168 https://en.wikipedia.org/wiki/Cinematograph#/media/File:CinematographeProjection.png

169 https:// en.wikipedia.org/wiki/35_mm_movie_film

170 http://worldkings.org/tag/hannibal-goodwin-patents-celluloid-photographic-film-(used-in-thomas-edison%27s-kinetoscope)-in-1887

171 https://Commons.Wikimedia.Org/Wiki/File:Cinnamomum_Camphora_-_K%C3%B6hler%E2%80%93s_Medizinal-Pflanzen-181.jpg

172 The Koishikawa Botanical Garden, which now belongs to Tokyo University, was originally established in 1684 as a herbarium by the Tokugawa Government and the Yojosho (a hospital free-for-charge for public) was added in 1722. The camphor tree is estimated as >300 year old. The picture was taken in January 2021 by the present writer.

173 https://item.rakuten.co.jp/mokuzai-o/kusu_chip_500/

174 Brett J. Stubbs. "Saviour to Scourge: a history of the introduction and spread of the camphor tree (Cinnamomum camphora) in eastern Australia". In: Brett J. Stubbs et al. (ed.), *Australia's Ever-changing Forests VI: Proceedings of the Eighth National Conference on Australian Forest History*. 2012. Original source: Donkin, R. A. "Dragon's Brain Perfume: an historical geography of camphor", Brill, Leiden 1999.

175 Ryoan Terashima, *Illustrated three elements of nature (heaven, earth and man) in Japan and China*, Vol.1 of 2, Chukindo Publ., 1884-1888 寺島良安(編)『倭漢三才圖會・下之巻』, 中近堂 (明治17-21).

176 Duncan, Robert Kennedy. *Some Chemical Problems of Today*, Harper & Brothers, New York, 1911.

177 J. J. Rein, *The Industries of Japan: Together with an account of its agriculture, forestry, arts, and commerce. From Travels and Researches Undertaken at the Cost of the Prussian Government.* A. C. Armstrong and Son, New York 1889.

178 Minoru Imoto, *The discovery of nylon/ The story of camphor*, Tokyo- Kagaku-dojin, 1971 (井本稔,『ナイロンの発見／樟脳ものがたり』, 東京化学同人,1971).

179 Ibid.

180 J. J. Rein, *The Industries of Japan: Together with an account of its agriculture, forestry, arts, and commerce. From Travels and Researches Undertaken at the Cost of the Prussian Government*, A. C. Armstrong and Son, New York 1889.

181 https://alchetron.com/Gustaf-Komppa.

182 George O. Poinar, *Life in Amber*, Stanford University Press, 1992, pp.5-15

183 Jean H. Langenheim, *Plant Resins: Chemistry, Evolution, Ecology and Ethnobotany, Chapter 4: Amber: Resins Through Geologic Time,* Timber Press 2003, pp.143-195

184 *Tentative method for determining the age of amber using DSC (TOPEM®),* Thermal Analysis Application No. UC 352, Application published in METTLER TOLEDO Thermal Analysis User Com 35. https://www.mt.com/vn/en/home/supportive_content/matchar_aps/MatChar_UC352.html

185 Feista, Michael; Lamprechtb, Ingolf; Müller, Frank. "Thermal investigations of amber and copal", *Thermochimica Acta 458*, 2007,162; Pagacz, Joanna; Stach, Paweł; Natkaniec-Nowak, Lucyn; Naglik, Beata; Drzewicz, Przemysław. "Preliminary thermal characterization of natural resins from different botanical sources and geological environments", *J. Thermal Analysis and Calorimetry 138(6)*, 2019, 4279– 4288.

186 Sold at amazon.co.jp.

187 https://balzekasmuseum.org/gift-shop/amber-cross-necklace/#.X9gJ_rPgpPZ

188 https://www.academia.edu/11254977/EBENA._Belgischer_Luxus_des_ Art_d%C3%A9co_aus_Kongo-Kopal_Objekte_der_K%C3%B6lsch_ Collection

189 Ibid.

[190] https://www.carters.com.au/index.cfm/item/1101666-a-pair-of- early-1900s-kauri-gum-busts-exquisitely-moulded-depict/Moko means Traditional tattoo of Maori people.

[191] https://www.invaluable.com/auction-lot/a-single-row-graduated-ambroid-bead-necklace-776-c-00e8d402c7

Index

Suffix after numbers. 'f': Figure legend, 't': Table.

* 9 7 8 4 9 0 9 6 0 1 8 4 1 *